Equity in the Urban Built Environment

This book explores inequities in the urban built environment across a diverse range of places and considers practical solutions and strategies aimed at building more just, inclusive, and sustainable cities.

Achieving more equitable and prosperous urban places requires a critical examination of the design and layout of our cities. The 16 chapters of this book illuminate the ways in which the built environment, including buildings, roads, public spaces, and other infrastructure, shapes our health and prosperity through a complex set of physical and social interactions. It brings together experts from a variety of fields to identify, and suggest workable solutions for, inequities in the spatial distribution of amenities and disamenities and the processes and policies that give rise to these unjust patterns. Blending scholarly knowledge and practical experience, many of these strategies and solutions are presented through the lens of real-world case studies. One key takeaway is that the planning of our cities should be a communal effort that properly reflects the needs of all residents. Equity in the built environment can only be realized when people of diverse backgrounds, ethnicities, ages, and abilities feel welcome to both shape and enjoy the shared public realm.

This book is intended for a wide audience that primarily includes scholars, students, and professionals in the fields of urban geography, urban planning, landscape architecture, and urban studies. Professionals in urban planning and policy with a desire to advance equity goals will likely appreciate the strategies, recommendations, and best practices discussed within.

Bradley Bereitschaft is Professor of Geography at the University of Nebraska Omaha. His research focuses on the role of urban design and transportation in achieving equitable, healthy, and productive urban places.

Routledge Studies in Urbanism and the City

Equity in the Urban Built Environment
Bradley Bereitschaft

Urban Ethics as Research Agenda
Outlooks and Tensions on Multidisciplinary Debates
Raúl Acosta, Eveline Dürr, Moritz Ege, Ursula Prutsch, Clemens van Loyen and Gordon Winder

Heritage, Crafting Communities and Urban Transformation
Durga Puja Festival, Kolkata
Debapriya Chakrabarti

Uneven Real Estate Development in Romania at the Intersection of Deindustrialization and Financialization
Edited by Enikő Vincze, Ioana Florea and Manuel B. Aalbers

Uneven Real Estate Development in Romania at the Intersection of Deindustrialization and Financialization
Edited by Enikő Vincze, Ioana Florea and Manuel B. Aalbers

Imagining Urban Complexity
A Humanities Approach in Tropes, Media, and Genres
Frans-Willem Korsten and Anthony Albright

Mapping Legalities
Urbanisation, Law and Informal Work
Edited by Thomas Coggin and Roopa Madhav

For more information about this series, please visit https://www.routledge.com/Routledge-Studies-in-Urbanism-and-the-City/book-series/RSUC

Equity in the Urban Built Environment

Edited by Bradley Bereitschaft

Routledge
Taylor & Francis Group
LONDON AND NEW YORK

First published 2025
by Routledge
4 Park Square, Milton Park, Abingdon, Oxon OX14 4RN

and by Routledge
605 Third Avenue, New York, NY 10158

Routledge is an imprint of the Taylor & Francis Group, an informa business

British Library Cataloguing-in-Publication Data
A catalogue record for this book is available from the British Library

ISBN: 978-1-032-63106-6 (hbk)
ISBN: 978-1-032-63109-7 (pbk)
ISBN: 978-1-032-63111-0 (ebk)

DOI: 10.4324/9781032631110

Typeset in Times New Roman
by SPi Technologies India Pvt Ltd (Straive)

To all those who have suffered and fought injustice, and all those upon which a better future now precariously rests.

Contents

Illustrations

Figures

Tables

Contributors

Sandra Aguilar-Gomez Department of Economics, Universidad de los Andes

Bradley Bereitschaft Department of Geography/Geology, University of Nebraska Omaha

Magdalena Biernacka University of Lodz

Adrienne Cachelin Department of Geography, University of Utah

Senlin Chen School of Kinesiology, Louisiana State University

Victoria Derr Department of Applied Environmental Science, California State University Monterey Bay

Si Gao Department of Environmental Studies, California State University, Sacramento

Catherine Garoupa Central Valley Air Quality Coalition

Gabrielle Gonzales Department of Sociology, California State University, Sacramento

Elizabeth Guffey Department of Art History, Purchase College

Megan E. Hatch Maxine Goodman Levin School of Urban Affairs, Cleveland State University

Ann Heylighen Research[x]Design, Department of Architecture, KU Leuven

Douglas Houston School of Social Ecology, University of California, Irvine

Leah Joyner Department of Recreation, Parks, and Tourism Administration, California State University, Sacramento

Gregg Macey School of Law, University of California, Irvine

Dustin C. Major Maxine Goodman Levin School of Urban Affairs, Cleveland State University

Anvar Mukhamedjanov Department of Urban Planning and Design, The University of Hong Kong

Sarah Nelson Department of Geography/Geology, University of Nebraska Omaha

Amit Patel McCormack Graduate School of Policy and Global Studies, University of Massachusetts Boston

Jeannine M. Pearce School of Social Ecology, University of California, Irvine

James B. Pick School of Business & Society, University of Redlands

William Renel Touretteshero CIC

Avijit Sarkar School of Business & Society, University of Redlands

Amanda E. Staiano Pediatric Obesity and Health Behavior Laboratory, Pennington Biomedical Research Center, Louisiana State University

Wendy Steele Centre for Urban Research, RMIT University

Bo-sin Tang Department of Urban Planning and Design, The University of Hong Kong

Fahui Wang The Pinkie Gordon Lane Graduate School, Louisiana State University

Timothy F. Welch School of Architecture and Planning, University of Auckland

Jack Yochum Maxine Goodman Levin School of Urban Affairs, Cleveland State University

Yutian Zeng Department of Geography and Anthropology, Louisiana State University

Equity in the Urban Built Environment

An Introduction and Overview

Bradley Bereitschaft

Though perhaps seldom thought of as such, cities are essentially the *de facto* "habitat" of *Homo sapiens*. Indeed, as most urban textbooks written within the last few years will tell you, the majority of the world's 8+ billion people now reside in cities (a threshold crossed around 2007; Ritchie et al. 2024). This urban habitat is not only where most of us live, but it is also where we learn, create, collaborate, share, and generate most of the ideas and much of the wealth upon which our modern society is built. The synergistic power of cities plays out on a complex and uneven stage composed of buildings, roads, utility infrastructure, and public spaces – the physical built environment. Far from passive or impartial, the built environment influences nearly every aspect of our lives. Access to quality housing, schools and healthcare facilities, healthy food venues, robust public transit systems, and safe, clean green spaces, for example, can have profound impacts on our physical and mental health, as well as our future opportunities and prospects. In a perfectly equitable society, everyone – regardless of class, age, gender/sexuality, race/ethnicity, or ability – would have access to the physical places and resources they need to live a healthy, prosperous life. Yet, despite the many technological and social advances over the past century, it is broadly understood that this landscape remains far from even. This edited volume brings together experts in the fields of Geography, Planning, Architecture, and Urban Studies to explore these inequities across a diverse range of places and built environments, and consider practical solutions and strategies – many informed directly by the research and experience of the authors – aimed at building more just and sustainable cities for all.

With few exceptions, the physical built environment of most cities today is one that has been intentionally and meticulously orchestrated. Master plans, zoning codes, ordinances, regulations, and a complex web of procedures, permits, and other legal frameworks relentlessly – and often without notice – shape the world around us. Whether intentional or not, many of these decisions (often written in legal codes) have had the effect of privileging certain groups over others, frequently resulting in a physical environment with markedly unequal benefits and costs. Conventional zoning in the United States, for example, has not only contributed to urban sprawl but has also long reinforced patterns of racial and class segregation by limiting the supply and increasing the cost of

DOI: 10.4324/9781032631110-1

housing (Gray 2022). As Hatch et al. discuss in their chapter *Affordable Housing, Equity and Zoning*, wealthier households – at the detriment of everyone else – have been incredibly successful at using zoning and other regulations to keep affordable (and often denser multiunit) housing out of their neighborhoods. More sinister were the discriminatory practices of "redlining," racial steering, and racial covenants, which were outlawed in the United States six decades ago yet continue to perpetuate severe inequities. Perhaps, most notably, these practices contributed to a startling wealth gap between white and minority (particularly Black) households by making it much more difficult for people of color to purchase or upgrade homes (Shapiro et al. 2013). Over time this divide has had a cascading effect on minority neighborhoods, leading to a spiraling decline in amenities and services and a rise in unemployment, poverty, and crime (Powell and Porter 2023).

Perhaps more than any other city, Detroit, Michigan frequently serves as an exemplar of this process of discrimination, segregation, and decline leading to stark inequities in the urban built environment. Given its legacy as the "The Motor City" (Greater Detroit is still home to the "big three" U.S. automotive manufacturers: General Motors, Ford, and Chrysler), it is fitting that Timothy Welch, in his chapter *The Motor City's Evolution: How the Automotive Industry Fueled Detroit's Transport Inequities*, uses Detroit to explore inequities in another key component of the built environment: transportation. Overall, the transition away from public transportation in favor of private automobility in Detroit, and indeed much of the world, has placed additional burden on disadvantaged groups. In an urban environment built primarily around automobiles, those who cannot afford or are unable to operate private vehicles are likely to encounter significant difficulties reaching daily destinations including work, school, grocery stores, and healthcare facilities. Indeed, due to a combination of economic decline and a rise in automotive dependency, access to healthy food options and medical care may be severely restricted in some areas, resulting in food or medical "deserts." Growth in the demand for walkable, transit-oriented neighborhoods in the United States, and in many other cities of the Global North (i.e., "gentrification"), have also made it increasingly difficult to afford housing in amenity-rich and transit-accessible areas, a topic I discuss in my chapter *Equity in Neighborhood Walkability: Challenges and Opportunities*.

Automotive tailpipe emissions, and other sources of pollution in the urban environment, pose another significant challenge to equity. Disadvantaged groups are often disproportionally exposed to high levels of pollution from both "point sources" (i.e., power plants, industrial facilities) and "non-point sources" (i.e., cars, airplanes). In addition to the chemical toxicity of air, water, and soil pollution, low-income and other vulnerable groups frequently contend with higher levels of light and noise pollution (Casey et al. 2017; Nadybal et al. 2020). Studies have found that low-income and minority households often live in more polluted areas of the city, as they may provide one of the few pockets of affordable housing. Additionally, disadvantaged, and often segregated, areas of the city have historically been less able to politically resist the construction of new sources of pollution such as highways, industrial facilities, or landfills (Taylor 2014). In the United States, highway construction throughout the middle and late 20th century was used as an urban renewal tactic to clear away so-called "blighted" central city areas. This process either displaced or destroyed hundreds of established neighborhoods and introduced a significant source of pollution that, decades later, continues to undermine the health and well-being of urban residents (Reft 2023). Using a case study of Fresno, California, Douglas Houston et al. consider the causes, impacts, and potential means of mitigating these inequities in their chapter *Near-Roadway Pollution in South Central Fresno, California: Can State-led, Place-based Community Engagement Advance Environmental Justice?*

Leah Joyner et al. continue the discussion of environmental inequities in the built environment with an examination of environmental justice at the intersection of food security in Sacramento, California. In their chapter *Understanding Environmental and Food (in)Justice through Community-based Research: Stories from Sacramento Community Coalitions*, the authors share results from a series of interviews with community members, which highlight the challenges and opportunities of urban agriculture amid disinvestment, environmental pollution, and threats posed by climate change. In addition to providing space for community gardens and urban agriculture, green spaces may offer residents numerous benefits including access to nature, space for recreation, and critical ecological services, a topic explored in detail by Magdalena Biernacka in her chapter *Green Space Provision and Equity in the City* and Bo-sin Tang and Anvar Mukhamedjanov in their chapter *Public Space and Environmental Justice in High-density Cities*. Including case studies primarily from Poland and Hong Kong, respectively, Biernacka and Tang/Mukhamedjanov explore the benefits and challenges of equitable green/blue space provision in the urban context.

When addressing environmental threats and inequities, it is crucial to have as much information as possible. In her chapter *Addressing Environmental Quality Inequities: The Role of Information and Collective Action*, Sandra Anguilar-Gomez provides an overview of the types of environmental information and methods of collection typically used to identify, assess, and monitor pollution (particularly air pollution) in the urban environment. While this process often includes more traditional ground- or satellite-based monitoring sensors operated by government agencies, citizen science programs and crowd-sourced information can also be deployed effectively to address environmental concerns as well as educate and engage the local community. As Anguilar-Gomez suggests, these diverse sources of information can then be used to better inform both individual and collective action.

Community engagement in the assessment and planning of urban built spaces is a key component of ensuring environmental (as well as economic and social) equity. In her chapter *Locating Representational Equity: Intersections between Child-Friendly Cities and Children's Roles in Realizing Environmental*

Justice, Victoria Derr uses examples from the United States and Italy to showcase how participatory planning efforts can improve the quality of life and reduce environmental health disparities among one of the city's most vulnerable populations: children. Zeng et al. further explore the relationships between the urban built environment and children's health in their chapter *Disparities in Obesogenic Built Environment for Children: A Pilot Study in East Baton Rouge Parish, Louisiana*. As with all urban residents, children's health is shaped by the physical environment in numerous ways. Zeng et al., in a case study of Baton Rouge, Louisiana, focus on two key attributes: the availability of spaces for physical activity (e.g., parks, recreational facilities) and the presence of retailers offering healthy food options (i.e., the "food environment"). Uncovering and addressing inequities in the built environment as it impacts the physical and mental well-being of children is particularly crucial to advancing long-term social justice as poor health can limit lifelong opportunities for upward mobility and perpetuate cycles of poverty (Manduca and Sampson 2019).

Advancing equity in the urban built environment by developing more age-friendly cities must also involve consideration of the unique needs of older residents. Sarah Nelson, in her chapter, *Health and Equity in Age-Friendly Cities and Communities: A Global Geographic Perspective*, again examines equity in health and well-being, but in the context of older populations. The author echoes the previous recommendation for community input (i.e., procedural equity) to respect cultural differences and ensure that place-specific needs are met. As older adults are likely to experience mobility constraints, the availability and physical accessibility of public transit, walkable neighborhoods and pedestrian pathways, and social, medical, and recreational amenities are no doubt crucial. Yet, as Nelson suggests, age-friendly planning should involve consideration of emotional and social landscapes, that is, the "person-environment relationship," as well as physical infrastructure.

The next two chapters further consider the challenges, strategies, and rewards of building more inclusive urban environments, particularly for those with varying sensory and mobility capabilities. In their chapter *On Equity and Designing Space for the Range of Human Ability*, Ann Heylighen and Elizabeth Guffey discuss the history of inclusivity and equity in the design of public spaces, including concepts such as inclusive design (ID) and universal design (UD). The authors share results from an enlightening interview with members of a Seattle-based ID/UD architectural consultancy firm. While there is clearly a pressing need to design for a range of human abilities and, in doing so, make every public space as welcoming as possible, the authors note that this process is not perfect. Improved access or navigation for some, for example, might in fact lead to a worse experience, or loss of access, for others. In such cases, a compromise or "happy medium" may offer the best overall outcome.

While discussions of equity in the built environment as it pertains to ID often focus on how best to accommodate and support those with differing mobility or visual capabilities, variations in the other human senses – including hearing – are less frequently considered. With this deficit in mind, William

Renel, in his chapter *Auditory Ableism: Exploring Aural Diversity, Equity and Belonging in the Built Environment*, considers how the diversity of human hearing (i.e., aural diversity) can pose unique challenges for many given the tendency of the built environment to cater primarily, or even solely, to the "auditory normate" (i.e., persons with normal hearing capabilities). Using a series of interviews and "soundwalks" (i.e., walking along with participants as they navigate public spaces), Renel explores how aural diversity and urban design combine to shape feelings of equity and inclusion.

The final three chapters consider a diverse group of contemporary challenges to equity in the urban built environment: accessibility of high-speed broadband internet and communications infrastructure, health emergencies like COVID-19, and climate change. Avijit Sarkar and James Pick discuss disparities in internet access, their causes, and implications for both individuals and society in their chapter *Digital Equity in the United States: Disparities and Socioeconomic Influences on Internet Access and Use*. In addition to differences in the availability and accessibility of internet-related infrastructure and services, the authors note that there remains a digital divide among various socioeconomic and demographic groups, perpetuated in part by disparities in education, skillsets, and social capital. Next, Amit Patel investigates the role of the built environment in both the spread and mitigation of diseases, particularly infectious diseases such as COVID-19, in his chapter *Impact of the Urban Built Environment on Public Health During the COVID-19 Pandemic*. Patel focuses primarily on the challenges faced by informal settlements and cities of the Global South, which often lack the physical and financial resources to effectively address public health emergencies. Lastly, Wendy Steele, in her chapter *Regenerative Futures: Climate Justice and Equity in the Built Environment* confronts one of the greatest threats to urban equity and sustainability in the years and decades to come. Although they are hotspots of energy consumption and waste generation, cities are also centers of innovation, collaboration, and efficient infrastructure (e.g., public transportation, multiunit housing) that can be leveraged to both mitigate and adapt to a rapidly warming planet. These physical and social resources can also be put into action to combat climate injustices, particularly where disadvantaged groups are disproportionally vulnerable to severe climate-related stresses and events.

Home to some 4.5 billion people and counting (Ritchie et al. 2024), our cities present many of the most significant opportunities for advancing equity in the 21st century. A critical consideration of the design and layout of our urban places – particularly public spaces – represents an integral part of that process. Equity in the spatial distribution of amenities (e.g., green space, quality schools, transit stations) and disamenities (e.g., landfills, busy roadways) (i.e., "distributional equity") and the processes and policies that give rise to these patterns (i.e., "procedural equity") must both be addressed. Crucially, the planning of our cities needs to be a communal effort that properly reflects the needs of all residents, not just the wealthy and powerful. Equity in the built environment can only be achieved when people of diverse backgrounds, ethnicities, ages, and abilities feel welcome in both helping to shape as well as enjoy the shared urban realm. As you read through the 16 chapters of this volume, I challenge you to consider inequities in the built environment you have encountered in your own city or community. How might these inequities be addressed and what role could you play? Hopefully this volume will give you some ideas. Even small adjustments in the service of greater equity and justice in the urban built environment can have far-reaching consequences. Let each of us endeavor to create more fair, just, and welcoming urban places – *habitat de homo sapiens* – for ourselves and all those who come after.

References

Casey, Joan A., Rachel Morello-Frosch, Daniel J. Mennitt, Kurt Fristrup, Elizabeth L. Ogburn, and Peter James, "Race/ethnicity, socioeconomic status, residential segregation, and spatial variation in noise exposure in the contiguous United States," *Environmental Health Perspectives* 125, no. 7 (July 2017): 077017, https://doi.org/10.1289/EHP89

Gray, Nolan. *Arbitrary Lines: How Zoning Broke the American City and How to Fix It*. Washington, D.C: Island Press, 2022.

Manduca, Robert, and Robert J. Sampson, "Punishing and toxic neighborhood environments independently predict the intergenerational social mobility of black and white children," *PNAS* 116, no. 16 (April 2019): 7772–7777, https://doi.org/10.1073/pnas.1820464116

Nadybal, Shawna M., Timothy W. Collins, and Sara E. Grineski, "Light pollution inequities in the continental United States: A distributive environmental justice analysis," *Environmental Research* 189 (October 2020): 109959, https://doi.org/10.1016/j.envres.2020.109959

Powell, Richard, and Jeremy Porter, "Redlining, concentrated disadvantage, and crime: The effects of discriminatory government policies on urban violent crime," *American Journal of Criminal Justice* 48 (October 2023): 1132–1156, https://doi.org/10.1007/s12103-022-09688-3

Reft, Ryan, Amanda Phillips de Lucas, and Rebecca Retzlaff, eds., *Justice and the Interstates: The Racist Truth about Urban Highways*. Washington, D.C.: Island Press, 2023.

Ritchie, Hannah, Veronika Samborska, and Max Roser (2024). "*Urbanization*." Accessed May 10, 2024. https://ourworldindata.org/urbanization

Shapiro, Thomas, Tatjana Meschede, and Sam Orso. 2013. *The Roots of the Widening Racial Wealth Gap: Explaining the Black-White Economic Divide*. Minority Health and Health Equity Archive. Research and Policy Brief #1. Waltham, MA: Institute on Assets and Social Policy, Brandeis University. Accessed May 21, 2024. http://iasp.brandeis.edu/pdfs/Author/shapiro-thomas-m/racialwealthgapbrief.pdf

Taylor, Dorceta. *Toxic Communities: Environmental Racism, Industrial Pollution, and Residential Mobility*. New York: New York University Press, 2014. https://doi.org/10.18574/nyu/9781479805150.001.0001

1 Affordable Housing, Equity, and Zoning

Megan E. Hatch, Dustin C. Major, and Jack Yochum

Introduction

Housing policy in general, and policies related to housing affordability specifically, have a history in the United States of being inequitable. Redlining, segregation, urban renewal, the foreclosure crisis, and evictions all disproportionately negatively impact(ed) people who are Black and/or have low incomes (among other groups). While all these areas are worthy of study, here we focus on how one policy – zoning – affects housing affordability inequity.

Zoning is a local government function that began in the United States in the late 1800s, with New York City implementing the first citywide comprehensive zoning law in 1916. Zoning regulates land use (residential, commercial, or industrial), density, site layout, and design (Ohio Department of Health 2019). Often, local governments employ zoning to limit the supply of housing (Brouwer and Trounstine 2024), which in turn can increase housing prices (Gyourko and Molloy 2015). In response to the growing affordability crisis, cities across the country initiated zoning reforms, with Minneapolis perhaps being the most well-known example. Evaluations indicate that the city's reforms led to more mid-sized apartment buildings – 87% of new housing between 2017 and 2022 were apartment buildings of 20 or more units – and kept rents from increasing as much as they increased throughout the rest of the state (Liang, Staveski, and Horowitz 2024; Wamsley 2024). Yet, other studies find zoning reforms primarily increase the supply of higher-cost housing units (Stacy et al. 2023), raising questions about the effectiveness of zoning reforms alone in addressing housing affordability.

In 2021, 21.6 million households in the United States were cost-burdened, meaning they spent more than 30% of their income on housing (Joint Center for Housing Studies of Harvard University 2023). Renters, in particular, were more likely to be cost burdened: 49% vs. 23% of homeowners (Joint Center for Housing Studies of Harvard University 2023). Nationwide, there are "only 34 affordable and available homes for every 100 extremely low-income renter households" (National Low Income Housing Coalition 2024, 1). While the reasons for inequitable housing affordability are multiple and complex, in this chapter, we use a four-part definition of equity (Johnson and Svara 2011; Svara

DOI: 10.4324/9781032631110-2

and Brunet 2004, 2005) to demonstrate the ways zoning creates and reinforces housing unaffordability and neighborhood exclusion by income, therefore increasing inequities in affordable housing provision in the United States.

The Four-Part Social Equity Framework

We use the four-part framework popularized by Svara and co-authors (Johnson and Svara 2011; Svara and Brunet 2004, 2005) to define social equity. According to this framework, social equity consists of *procedural fairness* (due process equity), *access* (distributional equity), *quality* (process equity), and *outcomes* (impact equity). Stokan et al. (2023) extend this framework to take into consideration the spatial dimensions of social equity, arguing that for some place-based policies, it is important to consider equity of access, procedural fairness, quality, and outcomes between neighborhoods or other geographic designations. Zoning is inherently spatial, and therefore, while we focus on how it affects the four dimensions of equity for different income groups, we also show how local zoning policies create large variations across jurisdictions as well as between neighborhoods within a jurisdiction, raising both intra- and intergovernmental equity concerns.

Procedural Fairness: Who Participates in the Zoning Process

The first dimension of social equity is procedural fairness, or due process in policy decision-making. In this context, procedural fairness is the "who" of zoning and how they participate in local government zoning deliberations. Land use decisions are inherently political, with deeply invested actors existing on all sides (Brouwer and Trounstine 2024). Equity planning emphasizes participation (Arroyo, Sandoval, and Bernstein 2023), often leaving to chance the question of who participates and how.

Local governments typically seek public input before making zoning decisions through public zoning board meetings. These meetings encourage members of the public to offer commentary about issues ranging from site projects to citywide reforms. While meetings are open to the public, concerns arise around individuals who are outspoken at these meetings: in addition to the lack of representation seen across public processes, small numbers of identified "neighborhood defenders" have a disproportionate impact on land use decisions (Einstein, Glick, and Palmer 2019). These neighborhood defenders are seldom representative of the broader community: they normally have higher socioeconomic statuses and significantly more antidevelopment motives than the rest of the community. These individuals may strongly identify with the Not In My Backyard (NIMBY) movement – a coalition of local actors that mobilize to ensure local planning decisions do not negatively impact their properties, often taking the form of advocacy against low-income housing (Ramzanpour, Sharghi, and Nourtaghani 2023). Counteracting NIMBYism is a growing contingency of Yes In My Backyard (YIMBY) actors in progressive urban areas who attempt to encourage development in positive ways, inadvertently allowing gentrification (Wyly 2022) or reinforcing segregation and disinvestment in neighborhoods (Curran 2022). Both groups influence development through participation in zoning decisions, often drowning out their opposition.

Traditional public meetings often fail to be representative of their communities across a spectrum of metrics. Factors such as timing, location, childcare availability, and social exclusion come together to deter participation from sizable portions of the population. Meeting content also determines who will attend. Williamson (2014), in her study of public meetings to decide how to spend block grant dollars that are primarily intended for the benefit of low- to moderate-income households, finds meeting participants hold more extreme views than the general public, but participants are more likely to be Black and have lower incomes, two groups that generally are observed to not attend public meetings. This instance is indicative of the fiefdoms residents perceive within their communities: those who are most immediately impacted by a decision are most likely to participate in public zoning meetings. Additionally, the procedures for these meetings are heavily biased toward existing residents (Einstein, Glick, and Palmer 2019), which leads to the development of localism or an intention to protect existing communities (Manville and Monkkonen 2024). People who would like to live in a new development in a community, but who are not yet residents, are very unlikely to attend a zoning board meeting about the proposed development, and even if they do attend, are unlikely to have standing to speak. In these instances, a lack of procedural fairness may lead to access inequities.

With the understanding that zoning board meetings are insufficient in capturing the input of the community, and in parallel with technological development of the early 21st century, alternative modes of public participation in zoning decisions have emerged. For example, participatory mapping is the process wherein selected community members construct zoning maps based on their knowledge of the pathways, strengths, opportunities, and challenges in the neighborhood (Brown, Sanders, and Reed 2018). The exact participatory nature is obscure: households need to be deliberately sampled in such a way that they are both geographically and sociodemographically representative; otherwise, existing and new spatial inequities risk being enforced (Brown and Kyttä 2014). Another innovation, community surveys, are also vulnerable to sampling issues but can be distributed both in paper and in digital formats, thus reaching a broader segment of the population than participatory mapping. These methods produce very similar results and are substantially less susceptible to outspoken actors than zoning board meetings (Brown and Eckold 2020).

Even when residents who have been historically excluded from zoning decision-making processes are included by zoning boards, procedural fairness concerns arise around how they are included. Predatory inclusion, first defined by Taylor (2019), refers to the trend where desired goods and services are given to traditionally excluded sociodemographic communities under stricter conditions than those extended to traditionally included communities. In the context of zoning, predatory inclusion describes the trend where disadvantaged populations are used as a means to an end in the decision-making process. Their voices serve as a checked box rather than a meaningful component of the plan. As with all advocacy efforts, planners run the risk of speaking on behalf of these populations rather than elevating the existing voices in the community (Hollander and Németh 2011). Therefore, zoning boards must be cognizant of their advocacy efforts for underserved populations to ensure procedural fairness. Without careful rethinking as to how, where, and when voices are solicited through zoning board meetings, local governments will continue to segregate neighborhoods and deprive historically underserved populations of equal access to high-quality housing in preferred neighborhoods.

Access: Who Can Live Where

Access equity refers to the distribution of households of various income levels within and between communities. While originally envisioned as a land use planning method, zoning is often weaponized to limit access to particular neighborhoods. Zoning created a formal tool with which planners and the communities they serve can reinforce segregation (Trounstine 2018), creating winners and losers (Serkin 2020). De jure segregation was established long before local zoning became commonplace, yet "zoning is perhaps the most successful segregation mechanism ever devised" (Gray 2022, 4). While explicit racial zoning practices have long since been abolished, zoning continues to reinforce existing disparities by prohibiting the development of affordable and multifamily housing, thus limiting access to prosperous communities (Brennan, Peiffer, and Burrowes 2019).

Restrictive zoning regulations often lead to low-density residential areas with high housing costs. When high land and construction costs are unable to be spread across multiple units, as in high-density developments, the result is higher housing prices and lower rates of available housing (Gyourko and Molloy 2015). High housing costs benefit existing homeowners while decreasing the purchasing power of households with low incomes that might otherwise look to move to high-cost cities that provide better economic opportunities (Ikeda and Washington 2015). When zoning regulations restrict high-density developments, those areas also tend to be more racially segregated compared to areas with more progressive zoning ordinances (Rothwell and Massey 2009). In contrast, when zoning reform allows for greater density, the supply of affordable housing can increase. For example, San Diego's accessory dwelling unit (ADU) Bonus program, which allows for additional units on existing lots and incentivizes affordable units, led to permits for an additional 295 deed-restricted affordable ADUs in two years (Terner Center for Housing Innovation U.C. Berkeley 2023).

Zoning boards often reinforce segregation through lot size requirements and site specifications. Communities employ minimum lot size requirements to deter the construction of multifamily housing, a process that demonstrably excludes Black and Hispanic families from communities with strict limits (Lewyn 2022). Rather than bar development outright, communities also place extensive restrictive regulations on developers that have the effect of pushing out developers altogether. Expulsive zoning, a term coined by urban planner Yale Rabin (1989), is perhaps the most egregious practice. While many zoning boards cave to neighborhood defenders, some zoning boards act in a more deliberate and sinister way by directing environmentally hazardous construction projects such as landfills and chemical factories into neighborhoods of color in an effort to remove those racial groups from the city.

To combat expulsive zoning and similar practices, equity planners adopt inclusive zoning practices. With inclusive zoning policies, local governments require developers to set aside a certain number of affordable units in their developments (Ramakrishnan, Treskon, and Greene 2019). The success of this practice depends heavily on what policy tools zoning boards can leverage and whether inclusionary zoning is mandatory, although thoughtfully implemented programs increase the number of affordable housing units (Hamilton 2021). However, these practices come at a cost to developers, which occasionally manifests as higher market rate housing and less overall construction (Ramakrishnan, Treskon, and Greene 2019). Similarly, the increasing popularity of form-based zoning codes illustrates another attempt at the integration of various income groups. Rather than dividing land based on land use, form-based zoning prioritizes aesthetic form, combatting sprawl while incentivizing mixed-use development (Steuteville 2017). The issues surrounding form-based code intersect both procedural fairness and access: form-based code is simple to implement due to its ability to fast-track development that fits with the form of that part of the city, increasing access but also the need for deliberate and expansive community input to ensure procedural fairness (Tagtachian, Barefoot, and Harreveld 2019).

The result of inclusionary zoning practices is a pathway for low-income residents into stable and desirable neighborhoods – in other words, greater access equity. However, McFarlane (2019) cautions that zoning for mixed-income communities recreates poor doors – side or back entrances to buildings that property managers require lower-income residents use – at a neighborhood level. Not only are residents with low incomes noticeable, but this inclusion also reinforces existing race and class biases. This illustrates how predatory inclusion practices creep into development even when the aim is inclusion: the very act of allowing low- and moderate-income populations into a neighborhood is both costly and socially disincentivized.

Local governments' capacity to govern development in a neighborhood also regulates access to that neighborhood. While natural and political boundaries play a major part in zoning decisions, the choices made within those constraints are meaningful for creating healthy, equitable neighborhoods accessible to a variety of people.

Quality: Who Gets Adequate Housing in Preferred Neighborhoods

From an affordable housing standpoint, the equitable provision of high-quality housing means having access to affordable housing free of physical and health-related problems (Raymond, Wheeler, and Brown 2011). When zoning regulations restrict the supply of affordable housing (Brouwer and Trounstine 2024; Gyourko and Molloy 2015), the remaining housing, often in other communities or specific areas within a community, is likely to be of lower quality than market rate housing.

Across the United States, the age of the housing stock continues to increase (Joint Center for Housing Studies of Harvard University 2023), which is only exacerbated by restrictive density zoning regulations (Rothwell and Massey 2009). As restrictive density zoning regulations keep higher-density, multifamily units from being built in and around single-family housing, many more affordable multifamily buildings are instead located near highways and in areas with higher risks of air pollutants (Brennan, Peiffer, and Burrowes 2019). In addition to poor environmental conditions, a review of housing stock by the Joint Center for Housing Studies at Harvard University (2023) estimated in 2021 that there were roughly 9.5 million occupied homes that had severe structural deficiencies or were lacking basic features like plumbing, electricity, and water. According to that study, low-income households had an estimated $57 billion in needed repairs. Buffalo addressed both housing quality and affordability through their green code, a zoning-based reform focused on environmental sustainability. A major tenet of that reform was a focus on rehabilitating existing historic buildings, including converting legacy commercial buildings into housing, a policy that led to over 10,000 new apartments (Steuteville 2022).

Without rehabilitation, aging housing is likely to have housing quality concerns. Some literature investigates the distribution of inadequate housing (e.g., mice, mold, no or insufficient heat, incomplete kitchen facilities, etc.), showing Black households are nearly two and a half times more likely to be living in inadequate housing, while Hispanic households are around two times as likely to be living in inadequate housing, compared to White households (Evans and Chapman 2024; Raymond, Wheeler, and Brown 2011). These racial and ethnic disparities hold across income levels, with Black and Hispanic households of all income levels being nearly twice as likely to be living in inadequate housing compared to White households at the same income levels (Evans and Chapman 2024). In addition to the racial and ethnic disparities in obtaining adequate housing, socioeconomic status also plays an influential role in determining who receives adequate housing. Households earning less than $25,000 annually were five times more likely to live in unsafe or unhealthy housing in 2009 than households earning more than $75,000 annually (Raymond, Wheeler, and Brown 2011). These disparities in who gets adequate housing demonstrate the access gap that leads to the inequitable provision of affordable housing.

Outcomes: Who Gets Affordable Housing in Their Preferred Neighborhoods

Outcome equity asks if restrictive zoning regulations result in affordable housing being available in desirable neighborhoods. That is, what is the quality of neighborhoods, based on economic and social outcomes, where affordable housing is commonly available, and who has access to higher-quality communities?

The suburbanization of employment has occurred in many metropolitan areas as suburbs welcome warehouses and office buildings and their high tax revenues to their communities (Smith and Steiner 2001). Due to exclusionary zoning practices in many of these suburbs, housing costs near where jobs are relocating prohibit low- and moderate-income earners from affording to live where they work, creating a spatial mismatch (Brouwer and Trounstine 2024; Smith and Steiner 2001). Limited access to well-paying employment opportunities and high housing costs limit the short- and long-term upward economic mobility of low-income and Black households by restricting earning potential and making homeownership less attainable (Brennan, Peiffer, and Burrowes 2019; Brouwer and Trounstine 2024). Beyond economic mobility concerns, high housing costs associated with restrictive land use regulations limit access for low-income and Black households to high-quality public services offered in many suburbs and wealthier communities, such as better public schools, parks, and local infrastructure (Whittemore 2021).

Exclusionary zoning practices limit the ability of low-income and Black households to access neighborhoods that could allow them to realize improvement in their economic and noneconomic outcomes. Even in cities with areas zoned for multifamily housing development, the high demand for multifamily housing and limited available land for such developments drives up the cost of development, eliminating the possibility of apartments being built for low- and moderate-income households (Whittemore 2021). Some cities even engage in what Dan Immergluck (2022, 206) calls "displace-and-replace" redevelopment where older, affordable multifamily housing is replaced with expensive, luxury multifamily housing.

Literature is unclear on what can be done to ensure affordable housing is developed in a way that guarantees equitable outcomes. In a study of inclusionary zoning policies in the Baltimore-Washington region, Hamilton (2021) finds that in jurisdictions requiring affordable housing be included in new housing developments, overall housing prices for market rate renters and homeowners increased to offset the costs of building the unsubsidized affordable units. This upward price trend also led to rising costs for low- to moderate-income households moving into the community. Another study contends cities that pass reforms relaxing zoning restrictions realize significant increases in their housing supply, especially for rental units affordable to households with above median incomes (Stacy et al. 2023). However, other studies argue for solutions that increase density in areas where people want to live, suggesting that the increase in available housing can control housing costs. For example, Schuetz (2022) asserts zoning changes allowing for more diverse structure types and sizes can serve as a solution to building more homes in desirable neighborhoods.

While inclusionary zoning policies enacted at the local level can be ineffective at increasing the availability of affordable housing because developers seek less restrictive markets, similar policies implemented at the state or regional level have the potential to be more effective in producing affordable housing in desirable neighborhoods (Kontokosta 2014). For example, because of the enforcement of the New Jersey court decisions establishing the *Mount Laurel Doctrine*, an annualized rate of 2,736 deed-restricted affordable units per year have been produced across the state between 2015 and 2022 (Fair Share Housing Center 2023b).

Restrictive zoning regulations limit the availability of affordable housing for low-income and Black households in highly desirable neighborhoods. Loosening zoning laws to allow for more variety in density may result in an increase in the housing supply in these neighborhoods. However, these measures alone will not fix the issues relating to a lack of affordable housing for low-income households, nor will they ensure outcome equity.

Conclusion

Almost half of renters and a quarter of homeowners in the United States spend more than 30% of their income on housing, indicating this country has an affordability problem. The reasons for a lack of affordability are complex and involve both insufficient income and high house prices. One factor contributing to inequities in housing is zoning. Housing is the major source of wealth in the United States, which means people who own their homes have a personal stake in protecting their wealth against perceived changes in property values, which leads to status quo bias (Davis, Rong, and Huennekens 2023; Schuetz 2022). This protectionism manifests in a lack of *procedural fairness* in zoning decision-making: those who would benefit from affordable housing are left out of the decision to build it, and instead, a small, unrepresentative group of homeowners often dominate, slow down, and even stop development. The resulting zoning code limits who have *access* to neighborhoods by creating impediments to affordable housing such as low-density requirements. The *quality* of affordable housing is often lower than the quality of more expensive housing, adding additional financial and social burdens to the people who live in low-quality homes. Together, inequities in procedural fairness, access, and quality lead to unequal *outcomes* such as segregation and a lack of economic mobility. As Minjee Kim (2024) argues, addressing zoning is an equity imperative to address the race- and class-biased origins of zoning in the United States.

Zoning reform, such as inclusionary zoning, form-based code, and upzoning, can help with these equity concerns, but they cannot solve the affordable housing crisis alone. "While zoning reform can help accommodate the growing demand for housing…it is not a panacea for generating more housing, and certainly by itself is unlikely to generate housing affordable to low-income families" (Immergluck 2022, 131). The effectiveness of zoning deregulation depends on housing market costs and whether changes are measured in the short- or long-term (Freemark 2023; Monkkonen, Manville, and Lens 2024), and may be stymied by states that preempt affordable housing policies such as inclusionary zoning (Goodman and Hatch 2023).

Waiting for filtering – the process of housing becoming affordable naturally with time – is not a solution to affordable housing inequities. To the extent it even happens, filtering is gradual and depends on the housing market (Spader 2024). Scholars assert that zoning reform, to be effective, will need to be a combination of a ban on exclusionary zoning, affirmative requirements for inclusionary zoning, reduction or elimination of parking requirements, and additional housing subsidies (Liang, Staveski, and Horowitz 2024; Rothstein 2017; Stacy et al. 2023). A "Fair Share Act" may be one solution. As proposed by John Charles Boger (1993, 1574), such a federal law would require and incentivize municipalities, through federal funding, to provide their fair share "of racially and economically integrative integrated housing." The *Mount Laurel Doctrine*, for example, led to over 300 New Jersey municipalities being in compliance with producing their fair share of affordable units (Fair Share Housing Center 2023a). On the other side of the equation, an increase in housing vouchers can assist individuals with low incomes in affording housing (Schuetz 2022).

Any lasting and complete solution to affordable housing will be political. While housing has generally been a local function in the United States, this local nature allows for exclusion and interjurisdictional inequities. Additional involvement of states and the federal government is likely needed (Lens and Monkkonen 2016; Schuetz 2022) to overcome status quo bias and each jurisdiction's orientation toward current residents and maintaining their tax base. This will need to be coupled with coalition building (Michener 2020). Housing in the United States has always been inequitable, and historically public policy has made it worse for marginalized groups; however, there is much opportunity to affirmatively address housing affordability and other inequities.

References

Arroyo, John C., Gerard F. Sandoval, and Joanna Bernstein. 2023. "Sixty Years of Racial Equity Planning: Evolution of an Ethic." *Journal of the American Planning Association* 89 (4): 438–58. https://doi.org/10.1080/01944363.2022.2132986

Boger, John Charles. 1993. "Toward Ending Residential Segregation: A Fair Share Proposal for the next Reconstruction." *North Carolina Law Review* 71 (5): 1573–1618.

Brennan, Maya, Emily Peiffer, and Kimberly Burrowes. 2019. "How Zoning Shapes Our Lives." Housing Matters: An Urban Institute Initiative. June 12, 2019. https://housingmatters.urban.org/articles/how-zoning-shapes-our-lives

Brouwer, N. R., and Jessica Trounstine. 2024. "NIMBYs, YIMBYs, and the Politics of Land Use in American Cities." *Annual Review of Political Science* 27 (1): 165–184. https://doi.org/10.1146/annurev-polisci-041322-041133

Brown, Greg, and Henry Eckold. 2020. "An Evaluation of Public Participation Information for Land Use Decisions: Public Comment, Surveys, and Participatory Mapping." *Local Environment* 25 (2): 85–100. https://doi.org/10.1080/13549839.2019.1703660

Brown, Greg, and Marketta Kyttä. 2014. "Key Issues and Research Priorities for Public Participation GIS (PPGIS): A Synthesis Based on Empirical Research." *Applied Geography* 46 (January): 122–36. https://doi.org/10.1016/j.apgeog.2013.11.004

Brown, Greg, Sara Sanders, and Pat Reed. 2018. "Using Public Participatory Mapping to Inform General Land Use Planning and Zoning." *Landscape and Urban Planning* 177 (September): 64–74. https://doi.org/10.1016/j.landurbplan.2018.04.011

Curran, Winifred. 2022. "Saying 'Yes' to What?: YIMBY and Urban Redevelopment in Chicago." *International Journal of Urban and Regional Research* 46 (2): 296–300. https://doi.org/10.1111/1468-2427.13063

Davis, Jenna, Helena Rong, and Joseph Weil Huennekens. 2023. "Perceptions toward Upzoning: A Parcel-Level Analysis of Public Sentiments toward the Minneapolis 2040 Plan." *Journal of Planning Education and Research*, November. https://doi.org/10.1177/0739456X231205584

Einstein, Katherine Levine, David M. Glick, and Maxwell Palmer. 2019. *Neighborhood Defenders: Participatory Politics and America's Housing Crisis*. Cambridge, UK: Cambridge University Press. https://doi.org/10.1017/9781108769495

Evans, Megan, and Alexander Chapman. 2024. "Residential Immobility and Racial and Ethnic Disparities in Housing Quality." *Sociology of Race and Ethnicity* 10 (1): 31–50. https://doi.org/10.1177/23326492231207603

Fair Share Housing Center 2023a. *Mount Laurel Doctrine Factsheet*. Cherry Hill, NJ: Fair Share Housing Center. https://www.fairsharehousing.org/wp-content/uploads/2024/08/Mount-Laurel-Doctrine_Fact-Sheet_Fair-Share-Housing-Center_July-2024.pdf

Fair Share Housing Center 2023b. "Dismantling Exclusionary Zoning: New Jersey's Blueprint for Overcoming Segregation - Fair Share Housing Center." April 25, 2023. https://www.fairsharehousing.org/dismantling-exclusionary-zoning-new-jerseys-blueprint-for-overcoming-segregation/

Freemark, Yonah. 2023. "Zoning Change: Upzonings, Downzonings, and Their Impacts on Residential Construction, Housing Costs, and Neighborhood Demographics." *Journal of Planning Literature* 38 (4): 548–70. https://doi.org/10.1177/08854122231166961

Goodman, Christopher B., and Megan E. Hatch. 2023. "State Preemption and Affordable Housing Policy." *Urban Studies* 60 (6): 1048–65. https://doi.org/10.1177/00420980221135410

Gray, M. Nolan. 2022. *Arbitrary Lines: How Zoning Broke the American City and How to Fix It*. Washington, DC: Island Press.

Gyourko, Joseph, and Raven Molloy. 2015. "Regulation and Housing Supply." In *Handbook of Regional and Urban Economics*, edited by Gilles Duranton, J. Vernon Henderson, and William C. Strange, 5:1289–1337. Elsevier. https://doi.org/10.1016/B978-0-444-59531-7.00019-3

Hamilton, Emily. 2021. "Inclusionary Zoning and Housing Market Outcomes." *Cityscape* 23 (1): 161–94.

Hollander, Justin B., and Jeremy Németh. 2011. "The Bounds of Smart Decline: A Foundational Theory for Planning Shrinking Cities." *Housing Policy Debate* 21 (3): 349–67. https://doi.org/10.1080/10511482.2011.585164

Ikeda, Sanford, and Emily Washington. 2015. *How Land-Use Regulation Undermines Affordable Housing*. SSRN Scholarly Paper. Rochester, NY. https://doi.org/10.2139/ssrn.3211656

Immergluck, Dan. 2022. *Red Hot City: Housing, Race, and Exclusion in Twenty-First Century Atlanta*. Oakland, CA: University of California Press.

Johnson, Norman J, and James H Svara. 2011. "Social Equity in American Society and Public Administration." In *Justice for All: Promoting Social Equity in Public Administration*, 3–25. M.E. Sharpe, Inc.

Joint Center for Housing Studies of Harvard University 2023. *The State of the Nation's Housing 2023*. Cambridge, MA: Harvard University. https://www.jchs.harvard.edu/sites/default/files/reports/files/Harvard_JCHS_The_State_of_the_Nations_Housing_2023.pdf

Kim, Minjee. 2024. "The Case for Mass Upzoning." *Housing Policy Debate* 34 (2): 246–51. https://doi.org/10.1080/10511482.2023.2234890

Kontokosta, Constantine E. 2014. "Mixed-Income Housing and Neighborhood Integration: Evidence from Inclusionary Zoning Programs." *Journal of Urban Affairs* 36 (4): 716–41. https://doi.org/10.1111/juaf.12068

Lens, Michael C., and Paavo Monkkonen. 2016. "Do Strict Land Use Regulations Make Metropolitan Areas More Segregated by Income?" *Journal of the American Planning Association* 82 (1): 6–21. https://doi.org/10.1080/01944363.2015.1111163

Lewyn, Michael. 2022. "Recent Case Law, Disparate Impact, and Restrictive Zoning." *Touro Law Review* 38 (2): 639–54

Liang, Linlin, Adam Staveski, and Alex Horowitz. 2024. "Minneapolis Land Use Reforms Offer a Blueprint for Housing Affordability." January 4, 2024. https://pew.org/3S76oXJ

Manville, Michael, and Paavo Monkkonen. 2024. "Unwanted Housing: Localism and Politics of Housing Development." *Journal of Planning Education and Research*, March. 44 (2): 685–700. https://doi.org/10.1177/0739456X21997903

McFarlane, Audrey G. 2019. "The Properties of Integration: Mixed-Income Housing as Discrimination Management." *UCLA Law Review* 66 (5): 1140–1215.

Michener, Jamila. 2020. "Power from the Margins: Grassroots Mobilization and Urban Expansions of Civil Legal Rights." *Urban Affairs Review* 56 (5): 1390–1422. https://doi.org/10.1177/1078087419855677

Monkkonen, Paavo, Michael Manville, and Michael Lens. 2024. "Built out Cities? A New Approach to Measuring Land Use Regulation." *Journal of Housing Economics* 63 (March): 101982. https://doi.org/10.1016/j.jhe.2024.101982

National Low Income Housing Coalition 2024. *The GAP: A Shortage of Affordable Homes, 2024*. Washington, DC: National Low Income Housing Coalition. https://nlihc.org/gap

Ohio Department of Health 2019. *Zoning 101: Frequently Asked Questions*. Columbus, OH: Ohio Department of Health. https://odh.ohio.gov/wps/wcm/connect/gov/791e1124-7fd6-4782-8392-367a68257391/FAQ_Factsheet_Land+Use+Zoning.pdf?MOD=AJPERES&CONVERT_TO=url&CACHEID=ROOTWORKSPACE.Z18_M1HGGIK0N0JO00QO9DDDDM3000-791e1124-7fd6-4782-8392-367a68257391-mJMzfuy

Rabin, Yale. 1989. "Expulsive Zoning: The Inequitable Legacy of Euclid." In *Zoning and the American Dream*, edited by Charles Monroe Haar. Chicago: Planners Press.

Ramakrishnan, Kriti, Mark Treskon, and Solomon Greene. 2019. *Inclusionary Zoning: What Does the Research Tell Us about the Effectiveness of Local Action?* Washington, DC: The Urban Institute.

Ramzanpour, Mehrnaz, Ali Sharghi, and Abdolmajid Nourtaghani. 2023. "Low-Income Housing Stigma, Results and Outcomes: A Systematic Review." *Journal of Housing and the Built Environment* 38 (2): 861–930. https://doi.org/10.1007/s10901-022-09967-9

Raymond, Jamie, William Wheeler, and Mary Jean Brown. 2011. "Inadequate and Unhealthy Housing, 2007 and 2009." In *CDC Health Disparities and Inequalities Report - United States, 2011 60*. Washington, DC: Centers for Disease Control. https://www.cdc.gov/mmwr/preview/mmwrhtml/su6001a4.htm

Rothstein, Richard. 2017. *The Color of Law: A Forgotten History of How Our Government Segregated America*. New York, NY: Liveright Publishing.

Rothwell, Jonathan, and Douglas S. Massey. 2009. "The Effect of Density Zoning on Racial Segregation in U.S. Urban Areas." *Urban Affairs Review* 44 (6): 779–806. https://doi.org/10.1177/1078087409334163

Schuetz, Jenny. 2022. *Fixer-Upper: How to Repair America's Broken Housing Systems*. Washington, DC: Brookings Institution Press.

Serkin, Christopher. 2020. "The Wicked Problem of Zoning." *Vanderbilt Law Review* 73 (6): 1879–1908.

Smith, Marc T., and Ruth L. Steiner. 2001. "Affordable Housing as an Adequate Public Facility." *Valparaiso University Law Review* 36 (2): 443–60.

Spader, Jonathan. 2024. "Has Housing Filtering Stalled? Heterogeneous Outcomes in the American Housing Survey, 1985–2021." *Housing Policy Debate*, 1–23. https://doi.org/10.1080/10511482.2023.2298256

Stacy, Christina, Chris Davis, Yonah Slifkin Freemark, Lydia Lo, Graham MacDonald, Vivian Zheng, and Rolf Pendall. 2023. "Land-Use Reforms and Housing Costs: Does Allowing for Increased Density Lead to Greater Affordability?" *Urban Studies* 60 (14): 2919–40. https://doi.org/10.1177/00420980231159500

Steuteville, Robert. 2017. "Great Idea: Form-Based Codes." *CNU*. May 10, 2017. https://doi.org/10/great-idea-form-based-codes

Steuteville, Robert. 2022. "How Zoning Reform Has Helped to Turn Buffalo Around." *CNU*. August 16, 2022. https://www.cnu.org/publicsquare/2022/08/16/how-zoning-reform-has-helped-turn-buffalo-around

Stokan, Eric, Megan E. Hatch, and Michael Overton. 2023. "Fifty Years as the Fourth Pillar of Public Administration: A Polycentric Extension of the Social Equity Framework." *Public Administration* 101 (4): 1427–42. https://doi.org/10.1111/padm.12888

Svara, James H., and James R. Brunet. 2004. "Filling in the Skeletal Pillar: Addressing Social Equity in Introductory Courses in Public Administration." *Journal of Public Affairs Education* 10 (2): 99–109. https://doi.org/10.1080/15236803.2004.12001351

Svara, James H., and James R. Brunet. 2005. "Social Equity Is a Pillar of Public Administration." *Journal of Public Affairs Education* 11 (3): 253–58. https://doi.org/10.1080/15236803.2005.12001398

Tagtachian, Daniela A, Natalie N Barefoot, and Adrienne L Harreveld. 2019. "Building by Right: Social Equity Implications of Transitioning to Form-Based Code." *Journal of Affordable Housing & Community Development Law* 28 (1): 71–115.

Taylor, Keeanga-Yamahtta. 2019. *Race for Profit: How Banks and the Real Estate Industry Undermined Black Homeownership*. Chapel Hill, NC: The University of North Carolina Press.

Terner Center for Housing Innovation U.C. Berkeley. 2023. "San Diego's Success in Spurring Missing Middle Housing: The Accessory Dwelling Unit Bonus Program." February 15, 2023. https://ternercenter.berkeley.edu/research-and-policy/san-diego-adu-bonus-program/

Trounstine, Jessica. 2018. *Segregation by Design: Local Politics and Inequality in American Cities*. New York, NY: Cambridge University Press.

Wamsley, Laurel. 2024. "The Hottest Trend in U.S. Cities? Changing Zoning Rules to Allow More Housing." *NPR*, February 17, 2024. https://www.npr.org/2024/02/17/1229867031/housing-shortage-zoning-reform-cities

Whittemore, Andrew H. 2021. "Exclusionary Zoning." *Journal of the American Planning Association* 87 (2): 167–80. https://doi.org/10.1080/01944363.2020.1828146

Williamson, Anne R. 2014. "Public Meetings as Sources of Citizen Input: Comparing Attendees with Citizens at Large." *The Social Science Journal* 51 (2): 191–200. https://doi.org/10.1016/j.soscij.2013.08.012

Wyly, Elvin. 2022. "YIMBY: The Latest Frontier of Gentrification." *International Journal of Urban and Regional Research* 46 (2): 319–30. https://doi.org/10.1111/1468-2427.13067

2 The Motor City's Evolution

How the Automotive Industry Fueled Detroit's Transport Inequities

Timothy F. Welch

The evolution of transportation systems plays a significant role in shaping urban inequity. Perhaps the best illustration of this is the case of Detroit, a city forever linked with the rise and fall of the automotive industry. The rapid ascension of the auto industry and the subsequent decline of public transportation in Detroit have contributed to entrenched social and economic disparities. A pattern of swift automobile dominance followed by a marked decrease in equity is particularly acute in the city. Still, similar patterns have occurred in thousands of cities across the United States and worldwide.

In the early 20th century, Detroit was a thriving metropolis fueled by the growth of the automobile sector and the promise of middle-class prosperity. The corners of the city were connected by networks of streetcars, providing affordable and accessible transportation to residents from all walks of life (Sugrue, 1996). However, as the automotive industry grew in prominence, Detroit's transportation landscape began to change, with the gradual dismantling of the streetcar system and the inevitable spread of highways and suburban sprawl. Quite rapidly, car ownership became a necessity rather than a luxury, and those who could not afford to own or maintain a vehicle found themselves increasingly isolated from economic and social opportunities (Farley et al., 2000).

The consequences of this shift have been profound and long-lasting, with significant implications for equity and social justice. Today, Detroit is a city marked by stark inequalities, with high levels of poverty, unemployment, and racial segregation. The legacy of car-centric development and a lack of investment in public transportation have left many neighborhoods disconnected from the rest of the city, with limited access to jobs, education, and essential services (Thomas, 2013). This experience is not unique to Detroit; cities across the United States and worldwide have grappled with the challenges of balancing economic growth and urban development with the need for equitable and sustainable transportation systems (Rothstein, 2017).

The concept of "transportation equity" has garnered more attention in recent years, emphasizing the importance of a fair and just distribution of the benefits and burdens of transportation systems (Litman, 2022). Advocates and policymakers have called for more significant investments in public transportation,

DOI: 10.4324/9781032631110-3

pedestrian and bicycle infrastructure, and transit-oriented development, focusing on the needs of low-income communities and communities of color (Karner et al., 2017). However, achieving transportation equity in practice is often elusive and made more complex by a constantly changing and refined set of measures.

This chapter examines the relationship between transportation and social equity in Detroit, tracing the city's metamorphosis from a thriving metropolis to a prime example of post-industrial decline – one that first echoed through the Rust Belt but has spread to nearly every modern city in many ways.

A key lesson is addressing the intersections of transportation, housing, and land use policies in creating more equitable and sustainable cities. The legacy of redlining and discriminatory housing policies in Detroit has had long-lasting impacts on the city's transportation landscape, with many neighborhoods that were historically denied access to home loans and other forms of investment now facing the greatest barriers to reliable and affordable transportation (Rothstein, 2017). Addressing these intersecting inequities requires a holistic approach that recognizes the interdependence of transportation, housing, and land use systems and works to promote more integrated and equitable planning processes.

In recent years, some promising examples of this kind of holistic planning have arisen in Detroit and other cities. For instance, the Detroit Collaborative Design Center (DCDC) at the University of Detroit Mercy has worked with community partners to develop neighborhood-based transportation plans that prioritize the needs of low-income and minority residents while also addressing broader issues of housing affordability, environmental justice, and economic development (O'Leary & Pitera, 2015). By taking a more comprehensive and community-driven approach to transportation planning, initiatives like these offer a glimpse of a more equitable and sustainable future for Detroit and beyond.

This is no small task, but there are signs of hope and progress in the growing movement for transportation justice and equity. Across the United States and worldwide, communities are coming together to demand more just and sustainable transportation systems, from the bus riders' unions of Los Angeles to the cycling activists of Bogotá (Lubitow et al., 2017; Torres-Barragan et al., 2020). Through their collective efforts, these movements are helping to shift the conversation around transportation equity and build the political will for change.

The Rise of the Automotive Industry in Detroit

Detroit's transformation into the "Motor City" is closely linked to the history of the automotive industry in the United States. From the early 20th century onward, Detroit emerged as the epicenter of automobile manufacturing thanks to its strategic location, abundant resources, and skilled workforce (Sugrue, 1996). The birth of the industry can be traced back to pioneers like Henry Ford and Ransom Olds, whose innovations revolutionized transportation and put Detroit on the map.

Ford's introduction of the Model T in 1908 and his innovative use of the assembly line made automobile ownership accessible to the masses for the first time. As production soared and the industry grew, Detroit became a magnet for automakers, with the "Big Three" – Ford, General Motors, and Chrysler – dominating the landscape by the 1920s. The automotive boom brought unprecedented prosperity to the city, creating a thriving middle class and cementing Detroit's status as a symbol of the American Dream (Sugrue, 1996). At one point, the opulence of Detroit's architecture, funded by the wealth generated by the auto industry, earned it the moniker "Paris of the Midwest" (Dennis, 2011).

The auto industry's impact on Detroit's economy and society cannot be overstated. At its peak in the 1950s, the industry employed over 350,000 workers in the city, with over 85,000 alone at Ford's Rouge factory, accounting for nearly a third of all jobs (Sugrue, 1996). High wages and generous benefits offered by the automakers helped to create a strong middle class, with many workers able to afford a mortgage, cars, and other trappings of the American Dream. However, this prosperity wasn't evenly distributed, and the industry's growth exacerbated the city's racial and economic inequalities.

African American workers faced significant barriers to entry into the auto industry, with many relegated to the lowest-paying and most dangerous jobs. In the early 20th century, most automakers refused to hire Black workers altogether, and those who did find employment in the industry often faced severe discrimination and harassment (Meier & Rudwick, 1979). It was not until the labor shortages of World War II that significant numbers of African Americans gained a foothold in the industry, and even then, they continued to face obstacles to advancement and equal treatment (Sugrue, 1996).

The rise of the automotive industry also profoundly impacted Detroit's urban development and transportation landscape, with significant implications for equity and social justice. The concept of "path dependence" (Sorensen, 2015) helps to explain how early decisions and investments in automobile infrastructure shaped the city's growth trajectory, creating a self-reinforcing cycle of car-centric development. As car ownership became more widespread, the city's once-extensive streetcar system began to decline, and the construction of highways and continually outward-growing suburbs facilitated the growth of sprawling, low-density development patterns (Farley et al., 2000).

"Spatial mismatch" (Kain, 1968) highlights how the decentralization of jobs and housing and limited public transit options created barriers to employment and economic mobility for low-income communities and communities of color. Studies have found that Detroit lags behind other metro areas in job accessibility by public transit. A 2011 Brookings Institution study determined that only 22% of Detroit's jobs were reachable via public transit within 90 minutes, ranking Detroit 36th out of 100 metro areas studied. This is compared to an average of 30% of jobs accessible by transit across the metro areas (Tomer et al., 2011). A 2012 analysis conducted at the University of Michigan similarly found that across 46 major metro areas, an average of 31.1% of jobs were accessible within 90 minutes by transit, while Detroit ranked near the bottom at 13.9% (Grengs, 2012).

The impact of this spatial mismatch is exemplified by stark disparities that emerged between Detroit's inner city and its suburbs post–World War II. While the city's population peaked at over 1.8 million in the 1950s, the subsequent decades saw a massive exodus of white residents to the suburbs, driven partly by federally subsidized mortgage programs and the construction of new highways (Farley et al., 2000). By the 1960s, Detroit had become a majority-Black city, with many remaining white residents concentrated in a few isolated enclaves (Sugrue, 1996).

The process of "white flight" and suburbanization had devastating consequences for Detroit's inner-city neighborhoods, left to grapple with a shrinking tax base, declining property values, and a lack of investment in basic services and infrastructure (Sugrue, 1996). At the same time, decentralization of jobs to the suburbs made it increasingly difficult for residents of these neighborhoods to access employment opportunities, as many lacked reliable transportation options (Farley et al., 2000).

The automotive industry's decline starting in the 1970s rippled throughout Detroit's economy, leading to the closure of many small businesses and the erosion of the city's middle class. The impact of this economic downturn wasn't evenly distributed. Low-income communities and communities of color suffered the brunt of the fallout. In many cases, these communities were hit hardest by spatial mismatch, and transportation inequities created the rapid adoption of the personal automobile, leaving residents particularly vulnerable to the effects of deindustrialization (Thomas, 2013).

The rise of the automotive industry in Detroit is a complex story with both positive and negative consequences. While it brought unprecedented prosperity and opportunity to the city, it also contributed to creating a car-centric transportation system that exacerbated existing inequalities and limited access to opportunities for those without automobiles. This pattern is not unique to Detroit; many cities across the United States and around the world have experienced similar challenges because of prioritizing private vehicle ownership over more equitable and sustainable transportation options (Glaeser & Kahn, 2004).

While cities grapple with the complex challenges of economic inequality and the need for more sustainable and resilient transportation systems, the lessons from Detroit's experiences with the automotive industry are more relevant than ever. By examining the historical forces that shaped the city's transportation landscape and the consequences of car-centric development, urban planners and policymakers can gain valuable insights into the importance of prioritizing equity and sustainability in transportation planning (Karner et al., 2017).

The Decline of Public Transportation in Detroit

As the automotive industry grew in prominence and Detroit became increasingly car-centric, the city's once-robust public transportation system deteriorated. The decline of public transit in Detroit was a gradual process shaped by a complex interplay of economic, political, and social factors, with significant consequences for equity and access to opportunity.

In the early 20th century, Detroit boasted one of the most extensive streetcar networks in the country, with over 500 miles of track connecting the city's neighborhoods and suburbs. The Detroit United Railway (DUR) system played a crucial role in the city's growth and development (Hyde, 1996). For many residents of Detroit, the streetcar was more than a means of transportation but a symbol of the city's prosperity and progress.

However, by the mid-20th century, the streetcar system was in decline, struggling to compete with the convenience and flexibility of the automobile. The decline of the streetcar system was partly fueled by the lobbying efforts of the automotive industry, which sought to promote car ownership and discourage public transportation use. The "Great American Streetcar Scandal" of the 1920s and 1930s involved a systematic campaign by General Motors and other automakers to replace streetcars with buses, which were seen as more flexible and cost-effective (Norton, 2011). This campaign played out in cities nationwide, including Detroit, with far-reaching consequences for public transit.

An example of the impact of this campaign is the Detroit Street Railways (DSR), which took over the city's streetcar system in 1922. In the 1930s, the DSR began to phase out streetcars in favor of buses, a decision influenced in part by the lobbying efforts of the automotive industry (Hyde, 1996). By the 1950s, the last streetcar line in Detroit had been shut down, and the city's once-extensive rail network had been largely dismantled (Farley et al., 2000).

The phaseout of Detroit's streetcar system in favor of buses by the mid-1950s was a consequential urban policy decision that disproportionately impacted Black communities. This shift was influenced by economic factors, including the growing dominance of the automobile industry and a broader national trend prioritizing highway construction and suburban development over public transit (Sugrue, 1996; Farley et al., 2000).

The decline of public transportation in Detroit exemplifies the concept of "transportation injustice," which emphasizes the disproportionate burdens and limited benefits experienced by marginalized communities (Martens et al., 2012). The consequences of Detroit's streetcar closures and reduced public transit services fell heavily on low-income and Black residents, exacerbating social and economic inequalities (Thomas, 2013). This uneven distribution of transportation access illustrates the importance of considering social equity in transportation policy and planning decisions (Karner & Niemeier, 2013). As Detroit grapples with its transportation history's legacies, prioritizing the needs and voices of its most vulnerable residents will be essential for creating a more just and equitable future (Grengs, 2012).

The lack of reliable, affordable, and accessible public transit options in Detroit has contributed to "transport disadvantage" (Lucas, 2012) and "social exclusion" (Preston & Rajé, 2007), limiting opportunities for employment, education, healthcare, and social participation. The story of James Robertson, a Detroit resident who gained national attention in 2015, illustrates the significant toll of transportation disadvantage. Robertson, a 56-year-old plastic injection molder, faced a grueling daily commute of 7 hours round trip, including a 21-mile walk, to reach his workplace in suburban Rochester Hills, which was not served by public transportation (Laitner, 2015). Robertson's experience underscores the profound challenges low-income residents face in navigating an inadequate public transit system without access to a car.

As Kunstler (1993) argues, and Robertson's commute exemplifies, car dependency creates two classes of people, those who have access to a car and those who do not, perpetuating a cycle of poverty and limited economic mobility for those without access to reliable transportation.

The decline of public transportation in Detroit has also had significant environmental and public health consequences. A reliance on private autos has contributed to high levels of air pollution, traffic congestion, and road fatalities, disproportionately impacting low-income communities and communities of color (American Lung Association, 2023; Singer, 2022). The lack of safe and accessible transportation options has also been linked to higher rates of obesity, diabetes, and other chronic health conditions in Detroit (Zenk et al., 2014).

Despite these challenges, there have been some efforts in recent years to revitalize Detroit's public transportation system. In 2017, the city launched the "ConnectTen" plan to improve bus service on ten key routes with higher frequencies and extended hours (Lewis, 2018). Additionally, in 2018, the city opened the "QLINE" streetcar, which runs along a 3.3-mile stretch of Woodward Avenue in downtown Detroit (Lewis & Ramirez, 2017). ConnectTen and the QLINE are the foundation of the city's broader vision for mobility innovation and equitable access for all Detroiters, as highlighted in the Streets for People plan, which prioritizes the development of a more inclusive, safe, and efficient public transportation system (City of Detroit, 2022).

As Detroit looks to the future, the RTA's vision for a more inclusive and efficient transportation network underscores the importance of regional consensus on funding strategies, prioritizing regional connectivity, community engagement, and integrated land use and transportation planning, which will be essential for creating a more just and sustainable transportation system in the long term (Regional Transit Authority of Southeast Michigan, n.d.; Transportation Riders United, 2021).

The Consequences of Car Dependency in Detroit

The automobile's dominance in Detroit's transportation landscape has had far-reaching consequences for the city's social and economic well-being, particularly in terms of equity and social justice. The car-centric development patterns that emerged in the wake of the automotive industry's rise have created a system in which access to reliable, affordable transportation is a prerequisite for full participation in the city's economic and social life.

This reliance on cars mirrors a broader national issue, where the high cost of car ownership creates significant barriers to workforce entry and societal participation, disproportionately affecting Black Americans and other communities of color. According to the National Equity Atlas, Black households are the least likely to have access to a vehicle, at 18%, highlighting disparities exacerbated by income and wealth gaps, racial segregation, and discriminatory pricing practices for auto loans and insurance (National Equity Atlas, n.d.).

In addition, average motorists may pay up to $650,000 over a lifetime to own a car, with society bearing over 40% of these costs, encompassing not only vehicle maintenance expenses but also societal costs such as infrastructure, public health, and environmental impacts. These expenditures place a disproportionate burden on the poorest Americans, for whom vehicle-related costs consume a significantly larger share of after-tax income compared to wealthier households, underscoring a transportation policy landscape that has historically prioritized personal vehicle use over more sustainable and equitable modes of transport (Gössling, Kees, & Litman, 2022). This inequitable system calls for a shift toward more inclusive transportation policies and infrastructure development to alleviate the economic burden on marginalized communities and ensure broader access to employment and social opportunities.

Car dependency exerts an enormous financial burden on households. A 2020 survey highlights the far-reaching economic impact of car ownership, revealing that approximately 43% of Americans have incurred debt due to car-related issues (LendingTree, 2020). The survey found that 28% of drivers would be unable to cover a $500 car repair without resorting to debt, with consumers of color and those impacted by the COVID-19 pandemic disproportionately affected (LendingTree, 2020). This financial vulnerability underscores the pressing need for accessible and affordable transportation alternatives, especially for low-income households and those impacted by economic downturns. As seen in Detroit, the high costs associated with car ownership can force difficult trade-offs and exacerbate existing inequalities, making the development of inclusive transportation solutions a critical priority.

The over, or complete, reliance on cars as the de facto mode of transport is often called "automobile dependency" (Newman & Kenworthy, 1999). The term has been in common use for the last two decades to describe how transportation systems and land use patterns can create a self-reinforcing cycle of car dependence with significant social, economic, and environmental costs. In Detroit, the prioritization of automobile infrastructure and the sprawling, low-density development patterns it enabled have created a transportation system heavily reliant on private car ownership, with limited alternatives for those who cannot afford or choose not to own a car.

More than half of Detroit residents live in areas with limited access to fresh and healthy food, a situation exacerbated by the lack of reliable transportation options. The term "food desert" describes areas where residents face significant barriers to accessing balanced sources of daily food availability, putting them at risk of premature illness and death (Mari Gallagher Research & Consulting Group, 2007). In 2004, this issue came to a head when the last major supermarket closed its doors, leaving the city without a single major grocery store within its 142.9 square miles (NPR, 2007). For residents without access to a car, the only options were to purchase lower-quality goods at a higher cost from convenience stores or undertake arduous multi-bus journeys requiring separate fares to neighboring counties.

However, a more nuanced approach reveals that the lack of car ownership often results in "food swamps," characterized by an abundance of fast-food outlets and convenience stores offering limited nutritious options (Taylor & Ard, 2015). This situation underscores the interconnectedness of transportation and food access inequities. Despite Detroit's vibrant urban agriculture movement and the presence of community gardens and farmers' markets, a substantial portion of the population, particularly in low-income neighborhoods, remains at a significant disadvantage due to the scarcity of accessible and affordable healthy food options. Addressing these disparities requires targeted policies and community-driven initiatives that recognize the interdependence of transportation, food access, and health outcomes.

The "right to the city" framework (Harvey, 2008) emphasizes the importance of equitable access to urban spaces, resources, and decision-making processes for all residents, particularly marginalized groups. In Detroit, the consequences of car dependency, such as the financial burden on low-income households and the isolation of communities underserved by public transit, highlight how transportation inequities can infringe upon residents' right to participate fully in the life of the city.

Car dependency in Detroit has exacerbated negative environmental and health impacts, especially for low-income communities and communities of color. The extensive presence of highways and parking lots contributes to significant air pollution, leading to severe health issues among nearby residents. Studies conducted by local organizations and researchers reveal that in Detroit, air pollution, including particulate matter and ozone, is responsible for an estimated 721 premature deaths annually, affecting disproportionately high numbers of residents in low-income and minority neighborhoods (Community Action to Promote Healthy Environments [CAPHE], n.d.). The primary sources of this pollution are automobiles, trucks, and construction equipment, underscoring the city's reliance on car transportation. Additionally, about 69,000 Detroit residents live within proximity to major freeways, and numerous schools are located near major roadways, placing children at elevated risk of respiratory problems, including asthma (Levin, 2023). This disparity is largely attributable to the concentration of polluting industries and transportation infrastructure in these communities, coupled with limited access to healthcare and resources to manage symptoms effectively.

Delray, a predominantly Latino neighborhood in southwest Detroit, finds itself at the complex intersection of transportation infrastructure, environmental justice, and public health. Major industrial facilities surround the neighborhood, including the Marathon Oil refinery, the Ambassador Bridge to Canada, and a web of highways and rail networks. As a result, Delray residents have long grappled with elevated rates of asthma, cancer, and other health issues, which they attribute to the concentration of polluting industries in their backyard (Restum, 2022; Allnutt, 2023; Trauth-Jurman, 2014).

Recent studies have shed light on the alarming state of air quality in Southwest Detroit, revealing that the area contends with some of the worst air pollution in the nation. This pollution takes a significant toll on health outcomes and deepens disparities within the community. For example, asthma hospitalization rates in Delray are a staggering three times higher than the state average, underlining a public health crisis that demands urgent attention (Restum, 2022). Furthermore, researchers have discovered dangerous levels of iron and other contaminants in road dust and sediment samples collected near industrial sites in Delray, providing tangible evidence of the health risks posed by industrial activities and the lack of effective measures to control fugitive dust (Allnutt, 2023).

The case of Delray resonates with findings from Los Angeles, where research indicates that neighborhoods with lower vehicle usage experience higher exposure to vehicular air pollution. This disparity is pronounced among tracts with higher non-White populations, with majority-White tracts contributing disproportionately to pollution in non-White areas. Historical decisions in freeway infrastructure and residential segregation have entrenched these disparities, suggesting systemic environmental injustices not limited to Delray but observed across cities like Los Angeles, emphasizing the need for inclusive urban and transport policies to address these inequities (Boeing, Lu, & Pilgram, 2023).

The story of Delray is just one example of how transportation policies and investments in Detroit have often prioritized the needs of industry and commerce over the health and well-being of marginalized communities. The "social determinants of health" framework (Marmot & Wilkinson, 2006) highlights how transportation access and other social and economic factors can influence health outcomes, with the negative impacts of car dependency in Detroit disproportionately affecting low-income communities and communities of color.

Efforts to address these disparities have often faced resistance from suburban communities, who have historically opposed regional transportation initiatives and other efforts to promote greater integration and equity. This resistance is not new, with a history of opposition to public transportation improvements tracing back to decisions made in the mid-20th century, significantly affecting the development and efficiency of Detroit's public transit system (Felton, 2014). The Regional Transit Authority (RTA), created in 2012 to oversee public transportation in the four-county Detroit metropolitan area, exemplifies contemporary efforts to navigate these historical and ongoing challenges. Despite support from business leaders, elected officials, and community advocates, the RTA has struggled to secure funding and political support, with suburban voters rejecting a ballot measure to fund the agency in 2016. This struggle highlights the complexities of regional transportation planning in an area with a rich yet tumultuous history of public transit (Detroit Regional Partnership, n.d.).

The lack of regional cooperation and investment in transportation continues to have significant consequences for low-income Detroiters, who often rely on an inadequate and unreliable public transit system. A Detroit People's Platform report reveals that one in three Detroiters do not own or share a car. This situation disproportionately affects Black and low-income residents, limiting their access to employment, educational opportunities, and essential services (Detroit People's Platform, n.d.). The advocacy work of the Detroit People's Platform underscores the urgent need for an equitable, reliable public transit system that meets the needs of all Detroit residents, especially those for whom public transit is a necessity.

As Detroit looks to the future, there is a growing recognition of the need to address the consequences of car dependency and promote more equitable and sustainable forms of transportation. In recent years, the city has made some progress in expanding public transportation options, such as the QLINE streetcar and the SMART bus system (Spangler, 2023). However, this progress comes after decades of neglect and disinvestment in public transit, a trend that began with the rise of the automotive industry and the prioritization of car-centric development, ultimately leading to the decline of Detroit's once-robust public transportation system.

Detroit must fundamentally shift its approach to transportation planning and investment decisions to achieve greater transport equity by prioritizing community engagement, equity, and sustainability. This shift will require confronting and addressing the legacy of racism and discrimination that has historically shaped Detroit's transportation landscape. It is only with the fundamental shifts away from the traditional planning approach that will move the city toward realizing a vision of a more just and inclusive city for all residents.

Lessons from Detroit: Strategies for Equitable Transportation

Detroit's experience with car dependency and its impact on urban inequality offers valuable lessons for cities seeking to create more equitable and sustainable transportation systems. The city's history highlights the urgent need for transportation policies that prioritize the needs of disadvantaged communities and promote social justice.

Investing in robust public transportation systems is one of the most critical strategies for advancing transportation equity. Reliable, affordable, and accessible public transit is essential for connecting residents to jobs, education, healthcare, and other opportunities, particularly in low-income communities and communities of color. As Chetty and Hendren (2015) found, access to public transportation is strongly associated with upward mobility, with children from low-income families who grew up in cities with better public transit earning significantly more as adults than those who grew up in cities with less access.

To ensure all people have a fair opportunity for advancement and to promote more equitable access to public transportation, Detroit and other cities must make a sustained commitment to funding transit operations and infrastructure. This requires dedicated revenue streams, such as sales taxes or congestion pricing, and policies prioritizing transit investment over highway expansion (TransitCenter, 2018).

Cities can also incentivize transit-oriented development and mixed-use neighborhoods, which can help reduce car dependency and promote walkability (Cervero, Ferrell, & Murphy, 2002; Zuk et al., 2015). Seattle's ORCA Lift program, which provides reduced fares for low-income riders, and the city's Transportation Equity Program, which focuses on improving transit service in underserved neighborhoods, offer examples of how cities can prioritize equity in public transportation (City of Seattle, 2022).

In addition to investing in public transportation, cities should prioritize the needs of pedestrians, bicyclists, and other vulnerable road users in their transportation planning and design. This includes investing in safe and accessible pedestrian and bicycle infrastructure, such as protected bike lanes or cycleways, wider sidewalks, and improved crosswalks like raised pedestrian crossings and curb extensions, particularly in low-income and minority neighborhoods that transportation planners have often neglected (Sandt et al., 2016).

Detroit has made some progress in this area in recent years with the development of the Joe Louis Greenway. This 27.5-mile recreational trail will connect neighborhoods across the city (City of Detroit, 2021). However, much more needs to be done to create a truly equitable and accessible transportation system that prioritizes the needs of all users, not just drivers.

Another critical strategy for promoting transportation equity is engaging communities in the planning and decision-making. Too often, transportation planning has been dominated by technical experts and political elites, with little input from the communities most impacted by these decisions (Karner & Niemeier, 2013). Community-based participatory planning (CBPP) offers a more inclusive and equitable approach in which residents actively shape the transportation policies and projects that affect their lives (Geekiyanage, Fernando, & Keraminiyage, 2021).

In Detroit, organizations like the Detroit People's Platform and the East Michigan Environmental Action Council have been working to amplify the voices of marginalized communities in transportation planning and advocacy. These groups have pushed for more equitable and sustainable transportation policies, such as expanding bus services, developing non-motorized infrastructure, and including community benefit agreements in transportation projects.

Cities can also leverage emerging technologies and innovative approaches to promote transportation equity. For example, shared mobility services like bike-share and scooter-share programs can help to expand access to affordable and convenient transportation options, particularly in underserved neighborhoods (Kodransky & Lewenstein, 2014). However, these programs must be designed with equity in mind, using strategies like income-based pricing, targeted outreach to low-income communities, and partnerships with community organizations (Creger et al., 2019).

In recent years, "mobility justice" has emerged as a framework for advancing transportation equity (Sheller, 2018). Mobility justice emphasizes the need for transportation systems that provide access to opportunities and address the broader social, economic, and environmental injustices that shape mobility patterns and experiences. This requires a holistic approach to address housing affordability, job access, and environmental health issues.

Achieving this vision will be difficult and, so far, has proved to be elusive for even the cities that are most actively pursuing equity. Realizing a future in which transportation serves as an equitable resource for all will demand a deep commitment and collaboration among a diverse array of stakeholders, from policymakers and planners to community organizers and advocates. The path forward depends on learning from the past, amplifying the voices of those most affected by transportation decisions, and embracing innovative strategies and approaches. While the challenges are significant, the vision of a more just and inclusive transportation system is one that Detroit and other cities must strive for. By grounding efforts in the principles of mobility justice, it is possible to work toward a future where transportation empowers rather than hinders, connects rather than divides, and uplifts the most vulnerable among us.

References

Allnutt, B. (2023). *Delray residents proved right by new study on toxic dust*. Outlier Media.

American Lung Association (2023). *State of the Air 2023*. https://www.lung.org/research/sota/city-rankings/states/michigan/wayne

Boeing, G., Lu, Y., & Pilgram, C. (2023). Local inequities in the relative production of and exposure to vehicular air pollution in Los Angeles. *Urban Studies*. https://doi.org/10.1177/00420980221145403

Cervero, R., Ferrell, C., & Murphy, S. (2002). *Transit-oriented development and joint development in the United States: A literature review*. Transportation Research Board.

Chetty, R., & Hendren, N. (2015). *The impacts of neighborhoods on intergenerational mobility: Childhood exposure effects and county-level estimates*. Harvard University. https://scholar.harvard.edu/files/hendren/files/nbhds_paper.pdf

City of Detroit (2021). *Joe Louis Greenway Framework Plan*. https://detroitmi.gov/sites/detroitmi.localhost/files/2022-04/JLG%20Framework%20Plan%20Vol1%20The%20Vision%20040822.pdf

City of Detroit (2022). *The City of Detroit Transportation Master Plan: Streets for People*. https://detroitmi.gov/sites/detroitmi.localhost/files/2022-09/Detroit_SFP_Plan_09142022_web_1.pdf

City of Seattle (2022). *Transportation Equity Program*. https://www.seattle.gov/transportation/projects-and-programs/programs/transportation-equity-program

Community Action to Promote Healthy Environments (n.d.). *Air pollutant sources, exposures, & health impacts*. University of Michigan School of Public Health. https://caphedetroit.sph.umich.edu/air-pollutant-source-exposures-health-impacts/

Creger, H., Espino, J., & Sanchez, A. S. (2019). *Autonomous vehicle heaven or hell? Creating a transportation revolution that benefits all*. The Greenlining Institute. https://greenlining.org/wp-content/uploads/2019/01/R4_AutonomousVehiclesReportSingle_2019_2.pdf

Dennis, E. P. (2011). Reconsideration of the rise and fall of Detroit through the lens of non-place (Master's thesis). University of Michigan. https://deepblue.lib.umich.edu/handle/2027.42/120372

Detroit People's Platform (n.d.). Mapping the road to transit justice in Detroit. https://www.detroitpeoplesplatform.org/groups/transit-justice-team/

Detroit Regional Partnership (n.d.). Transportation. https://www.detroitregionalpartnership.com/transportation/

Farley, R., Danziger, S., & Holzer, H. J. (2000). *Detroit divided.* Russell Sage Foundation.

Felton, R. (2014, March 11). How Detroit ended up with the worst public transit. *Detroit Metro Times*. https://www.metrotimes.com

Geekiyanage, D., Fernando, T., & Keraminiyage, K. (2021). Mapping participatory methods in the urban development process: A systematic review and case-based evidence analysis. *Sustainability*, 13(16), 8992. https://doi.org/10.3390/su13168992

Glaeser, E. L., & Kahn, M. E. (2004). Sprawl and urban growth. In J. V. Henderson and J. F. Thisse (Eds.), *Handbook of regional and urban economics* (Vol. 4, pp. 2481–2527). Elsevier.

Gössling, S., Kees, J., & Litman, T. (2022). The lifetime cost of driving a car. *Ecological Economics*, 194, 107335. https://doi.org/10.1016/j.ecolecon.2022.107335

Grengs, J. (2012). Equity and the social distribution of job accessibility in Detroit. *Environment and Planning B: Planning and Design*, 39(5), 785–800. https://doi.org/10.1068/b36097

Harvey, D. (2008). The right to the city. *New Left Review*, 53, 23–40.

Hyde, C. K. (1996). "Planning a Transportation System for Metropolitan Detroit in the Age of the Automobile": The Triumph of the Expressway. *Michigan Historical Review*, 22(1), 59–95. https://doi.org/10.2307/20173555

Kain, J. F. (1968). Housing segregation, negro employment, and metropolitan decentralization. *The Quarterly Journal of Economics*, 82(2), 175–197. https://doi.org/10.2307/1885893

Karner, A., Golub, A., Martens, K., & Robinson, G. (2017). Transportation and environmental justice: History and emerging practice. In S. Holifield, J. Chakraborty, & G. Walker (Eds.), *The Routledge handbook of environmental justice* (pp. 400–411). Routledge.

Karner, A., & Niemeier, D. (2013). Civil rights guidance and equity analysis methods for regional transportation plans: A critical review of literature and practice. *Journal of Transport Geography*, 33, 126–134. https://doi.org/10.1016/j.jtrangeo.2013.09.017

Kodransky, M., & Lewenstein, G. (2014). *Connecting low-income people to opportunity with shared mobility*. Institute for Transportation and Development Policy. https://www.itdp.org/wp-content/uploads/2014/10/Shared-Mobility_Full-Report.pdf

Kunstler, J. H. (1993). *The geography of nowhere: The rise and decline of America's man-made landscape*. Simon & Schuster.

Laitner, B. (2015, January 31). *Heart and sole: Detroiter walks 21 miles in work commute*. Detroit Free Press. https://www.freep.com/story/news/local/michigan/oakland/2015/01/31/detroit-commuting-troy-rochester-hills-smart-ddot-ubs-banker-woodward-buses-transit/22660785/

LendingTree (2020, July 28). 43% of Americans driven into debt over car troubles. https://www.lendingtree.com

Levin, K. (2023, October 17). A guide to Detroit air quality and its health impacts. *Outlier Media*. https://outliermedia.org/a-detroiters-guide-to-the-air-we-breathe

Lewis, S. (2018, August 28). Riding the bus in Detroit soon will be quicker during peak hours. *The Detroit News*. https://www.detroitnews.com/story/news/local/detroit-city/2018/08/28/new-ddot-connection-service-adds-more-trips/1122586002/

Lewis, S. D., & Ramirez, C. E. (2017, May 12). QLine revives Detroit's streetcar history. *The Detroit News*. https://www.detroitnews.com/story/news/local/detroit-city/2017/05/12/qline-debut-detroit/101573838/

Litman, T. (2022). *Evaluating transportation equity: Guidance for incorporating distributional impacts in transportation planning*. Victoria Transport Policy Institute.

Lubitow, A., Rainer, J., & Bassett, S. (2017). Exclusion and vulnerability on public transit: Experiences of transit dependent riders in Portland, Oregon. *Mobilities*, 12(6), 924–937. https://doi.org/10.1080/17450101.2016.1253816

Lucas, K. (2012). Transport and social exclusion: Where are we now? *Transport Policy*, 20, 105–113. https://doi.org/10.1016/j.tranpol.2012.01.013

Mari Gallagher Research & Consulting Group. (2007). *Examining the impact of food deserts on public health in Detroit*. https://www.marigallagher.com/2007/06/19/examining-the-impact-of-food-deserts-on-public-health-in-detroit-june-19-2007/

Marmot, M., & Wilkinson, R. (Eds.). (2006). *Social determinants of health*. Oxford University Press.

Martens, K., Golub, A., & Robinson, G. (2012). A justice-theoretic approach to the distribution of transportation benefits: Implications for transportation planning practice in the United States. *Transportation Research Part A: Policy and Practice*, 46(4), 684–695. https://doi.org/10.1016/j.tra.2012.01.004

Meier, A., & Rudwick, E. (1979). *Black Detroit and the rise of the UAW*. Oxford University Press.

National Equity Atlas (n.d.). Car access: Insights & analyses. *National Equity Atlas*. https://nationalequityatlas.org/indicators/Car_access

Newman, P., & Kenworthy, J. (1999). *Sustainability and cities: Overcoming automobile dependence*. Island Press.

Norton, P. D. (2011). *Fighting traffic: The dawn of the motor age in the American city*. MIT Press.

NPR (2007). Editorial: No more supermarkets?: Major grocers flee Detroit. *National Public Radio*. https://www.npr.org/templates/story/story.php?storyId=12477872

O'Leary, C., & Pitera, D. (2015). Moving forward together: Engagement in community design and development. In L. Abendroth & B. Bell (Eds.), *Public interest design practice guidebook* (pp. 109–114). Routledge.

Preston, J., & Rajé, F. (2007). Accessibility, mobility and transport-related social exclusion. *Journal of Transport Geography*, 15(3), 151–160. https://doi.org/10.1016/j.jtrangeo.2006.05.002

Regional Transit Authority of Southeast Michigan (n.d.). Projects. https://rtamichigan.org/improving-regional-transit/

Restum, A. (2022). *Air quality in Southwest Detroit: A public health crisis*. Harvard Medical School Center for Primary Care.

Rothstein, R. (2017). *The color of law: A forgotten history of how our government segregated America*. Liveright Publishing.

Sandt, L., Combs, T., & Cohn, J. (2016). *Pursuing equity in pedestrian and bicycle planning*. U.S. Department of Transportation. https://rosap.ntl.bts.gov/view/dot/50792

Sheller, M. (2018). *Mobility justice: The politics of movement in an age of extremes*. Verso Books.

Singer, J. (2022). *There are no accidents: The deadly rise of injury and disaster—who profits and who pays the price*. Simon and Schuster.

Sorensen, A. (2015). Taking path dependence seriously: An historical institutionalist research agenda in planning history. *Planning Perspectives*, 30(1), 17–38. https://doi.org/10.1080/02665433.2013.874299

Spangler, T. (2023, May 3). *Detroit's new transit center construction tied to Amazon project, creating a stir*. Detroit Free Press. https://www.freep.com/story/news/local/michigan/detroit/2023/05/03/detroit-transit-center-construction-amazon/70176874007/

Sugrue, T. J. (1996). *The origins of the urban crisis: Race and inequality in postwar Detroit*. Princeton University Press.

Taylor, D. E., & Ard, K. J. (2015). Food availability and the food desert frame in Detroit: An overview of the city's food system. *Environmental Practice*, *17*(2), 102–133.

Thomas, J. M. (2013). *Redevelopment and race: Planning a finer city in postwar Detroit*. Wayne State University Press.
Tomer, A., Kneebone, E., Puentes, R., & Berube, A. (2011). *Missed opportunity: Transit and jobs in metropolitan America*. Brookings Institution. https://www.brookings.edu/research/missed-opportunity-transit-and-jobs-in-metropolitan-america/
Torres-Barragan, C. A., Cottrill, C. D., & Beecroft, M. (2020). Spatial inequalities and media representation of cycling safety in Bogotá, Colombia. *Transportation Research Interdisciplinary Perspectives*, 7, 100208. https://doi.org/10.1016/j.trip.2020.100208
TransitCenter (2018). *Inclusive transit: Advancing equity through improved access & opportunity*. https://transitcenter.org/wp-content/uploads/2018/07/Inclusive-1-1.pdf
Transportation Riders United (2021, October 24). New RTA regional transit plan proposes strong vision, strategies, action steps. https://www.detroittransit.org
Trauth-Jurman, D. (2014). *The story of Delray: A case study on environmental and restorative justice in Detroit*. Bowling Green State University Honors Projects.
Zenk, S. N., Schulz, A. J., Matthews, S. A., Odoms-Young, A., Wilbur, J., Wegrzyn, L., ... & Stokes, C. (2014). Activity space environment and dietary and physical activity behaviors: A pilot study. *Health & Place*, 30, 67–72. https://doi.org/10.1016/j.healthplace.2014.07.005
Zuk, M., Bierbaum, A. H., Chapple, K., Gorska, K., Loukaitou-Sideris, A., Ong, P., & Thomas, T. (2015). *Gentrification, displacement and the role of public investment: A literature review*. Federal Reserve Bank of San Francisco.

3 Equity in Neighborhood Walkability

Challenges and Opportunities

Bradley Bereitschaft

"Live where you love" proclaims the popular online service Walk Score® (www.walkscore.com), which allows users to input an address or city and see where it falls on a walkability spectrum that ranges from zero – essentially no walkability – to 100, indicating a "walker's paradise." If you live in the United States, chances are that your own neighborhood is not particularly walkable. The average Walk Score® for cities of 200,000 or more in the United States is just 48 – a score low enough to be described as "car dependent." Indeed, the overwhelming trend in the spatial development of the American urban landscape over the last seven decades has been low-density, auto-dependent suburban diffusion (Nechyba and Walsh 2004). Although the United States may exemplify the phenomenon, suburban sprawl is by no means uncommon among the world's industrialized nations (Herzog 2015; Liu and Meng 2020). As decades of scholarly research have shown, this dramatic shift in the way we build, live within, and travel about cities has had a significant, and often deleterious, impact on our health, the environment, and society (Litman 2003; Bueno-Suárez and Coq-Huelva 2020). How then can we enhance walkability and decrease auto-dependency while also ensuring that walkable neighborhoods and cities are broadly accessible and inclusive? This chapter will speak to this challenge, highlighting recent scholarly and practical research on walkable urbanism; what it is, why it is beneficial, what strategies may be employed to enhance accessibility, and what barriers remain to be overcome.

What Is Walkability?

In the historical context of human civilization, automobility – "the social and technical system of the car" (Urry 1999) that includes all auto infrastructure and supportive industries – is a startingly new phenomenon, but one that has so thoroughly entrenched itself in the psyche and physical lived reality of billions that its absence, for many, is nearly unthinkable. Yet, prior to the mid-20th century, cities were built primarily around human locomotion with a well-connected street network and a density and mix of land uses that often made it practical, if not necessary, to travel by foot. Although horse- and later electric- and diesel-powered public transit systems increased mobility and

DOI: 10.4324/9781032631110-4

speed within cities (as they do today), these systems also inherently encourage walking as passengers first need to access communal stops or stations. Public transit, in turn, tends to facilitate additional density, making walking and cycling more viable (Lachapelle and Noland 2012). Today, urban development that attempts to replicate this traditional pedestrian-centric design is often referred to as "new urbanism" or "neo-traditional development," while focusing higher density residential and commercial development around transit (particularly fixed rail) stations is frequently described as "transit-oriented development" (TOD) (Knaap and Talen 2005; Knowles, Ferbrache, and Nikitas 2020).

Each of these strategies, and their various components, either directly or indirectly attempt to increase walkability, generally by bringing destinations closer together and improving the viability of automotive alternatives. However, if walkability is broadly defined as the propensity of a place to encourage walking for transportation, then it depends on more than density, street connectivity, or the mixing of residential and commercial land use. Though combining these "macro-scale" elements of urban form is crucial, other "micro-scale" features of the streetscape such as the availability and condition of the sidewalk, building setbacks and sense of enclosure, presence of shade trees, overhead lights, benches, and crosswalks, among others, are also expected to influence walking behavior (Ewing and Handy 2009; Lo 2009; Ewing et al. 2016). In addition to physically facilitating travel by foot, many of these microscale elements of the built environment may impact walking behavior by shaping pedestrians' sense of safety and comfort. Decades after Jane Jacobs wrote of the need for "eyes upon the street" (1961, p. 35), research has shown that pedestrians feel safer in areas with an active streetlife (e.g., dining, performances, etc.), ample storefront windows, and proper street lighting, which together provide "natural surveillance" (Park and Garcia 2020). Additionally, dangers posed by vehicular traffic can be moderated by providing buffers (e.g., parked cars, trees, bike lanes) between the pedestrian and the roadway, utilizing various techniques to slow traffic, and ensuring that crosswalks and crossing lights are in place and well-marked (Quistberg et al. 2015; Stoker et al. 2015; Kweon et al. 2021). Scholars have also noted that pedestrians prefer spaces that are interesting and aesthetically pleasing; poorly maintained or unkept streetscapes and buildings can reduce feelings of safety and security, while sensorially dull environments are less likely to encourage walking behavior (Mehta 2008; Johansson, Sternudd, and Karrholm 2016).

Why Walkability?

The potential benefits of walkable urban environments to both individuals and society are numerous. The bulk of scholarly research in this area has focused on the relationships between the built environment and outcomes that fall within four overlapping arenas: human health, the environment, economics, and social dynamics. Regarding health and walkability, the literature overwhelmingly suggests a positive correlation. Recent studies and meta-analyses point to strong connections between elements of walkable urban design and daily physical activity, as well as positive health outcomes including lower risk of obesity, cardiovascular disease, and certain types of cancers (India-Aldana et al. 2023; Makhlouf et al. 2023). Less auto-centric and more walkable urban typologies are also expected to result in fewer pedestrian collisions due to fewer vehicle miles traveled (VMT), slower automotive speeds, and more abundant pedestrian infrastructure (Dumbaugh and Li 2010; Yeo, Park, and Jang 2015). Recent research suggests that, at least in certain contexts, the protective effects of walkable urbanism may be strongest for the lowest socio-economic groups (Lang et al. 2022).

The environmental benefits of walkability rely primarily on shifting transportation modes away from private vehicles. The relationship is straightforward: as people walk more (or utilize other methods of active transportation or public transit), they drive less and consequently release fewer air pollutants, carbon dioxide, and tire-related microplastics into the environment (Kole et al. 2017; Glazener and Khreis 2019). Paradoxically, residents of more walkable neighborhoods – who are expected to drive less and generate less air pollution – may be exposed to disproportionately high levels of airborne toxins due to congestion in dense urban settings and the tendency of air to become trapped between multistory buildings – that is, the "street canyon effect" (Ng and Chau 2014; Bereitschaft 2015). While changes to building design and orientation could lessen these exposures (Ng and Chau 2014), such modifications to the built environment are often slow and must compete with a wide range of other considerations. Planting more pollutant-absorbing greenery, instituting congestion pricing, and reducing roadway areas in favor of cycling and pedestrian corridors are likely to result in more immediate improvements (Brugge et al. 2015; Huang et al. 2021).

Many of the economic gains from walkability are also likely to arise from improvements in public health and environmental quality. Lower rates of disease from enhanced physical activity and reduced environmental pollution, coupled with fewer automotive fatalities, are expected to result in significant cost savings – possibly millions of dollars for individual communities and hundreds of millions for cities (Litman 2003; Trubka, Newman, and Bilsborough 2010). Cities that build more densely and invest in public transit are also likely to reduce costs associated with automotive and other public infrastructure whose provision is fiscally inefficient at low densities (Burchell and Mukherji 2003). Similarly, household cost savings from not having to purchase, maintain, fuel, insure, or store private vehicles are likewise substantial and allow money to be spent elsewhere (Nahlik and Chester 2014). For many individuals and households, particularly those with low income, walkability and transit also provide access to education, jobs, and other economic opportunities that may otherwise be out of reach (Litman 2003).

Lastly, walkability is likely to benefit social capital as well by providing the means and spaces for interaction (Rogers, Gardner, and Carlson 2013). Lively sidewalks and other public areas including parks and plazas facilitate both planned and chance encounters (Leyden 2003). Mixed-use developments and an abundance of "third places" (i.e., places where socialization occurs outside of home or work; Oldenburg 1999) such as coffee shops, cafés, and bars provide additional spaces for interaction, as well as contribute to the vitality of local street life (Jacobs 1961; Mehta and Bosson 2010). Walkable and transit-accessible urban spaces allow residents to access third places and civic/community centers, as well as visit friends and family, without the need for an automobile (van den Berg, Sharmeen, and Weijs-Perrée 2017). This is particularly crucial for low-income households, the young, elderly, and disabled, who may not otherwise be able to maintain these supportive social networks (Levasseur et al. 2015). In addition to enhancing social inclusivity and equitability, there is mounting evidence that dense, walkable urban spaces can enhance innovation, especially within start-up ecosystems, by facilitating professional and personal social exchanges, collaboration, and knowledge spillovers (Fang and Rao 2021; Bereitschaft 2023).

Is Walkability Accessible?

Given the variety of benefits associated with walkable urban environments, it is worth considering whether such spaces are becoming increasingly exclusive. This concern has grown more salient over the past two decades due to widespread inner-city revitalization efforts and gentrification-led displacement of lower-income and other disadvantaged groups (Zuk et al. 2018). Access to walkable and transit-oriented environments may be especially crucial for disadvantaged groups, including low-wage workers, immigrants, and ethno-racial minorities, who often have lower rates of car ownership and are more likely to rely on alternative modes of transportation (Smart and Klein 2015). An obvious first question, then, is whether walkable neighborhoods are affordable for low- and moderate-income groups, and whether affordability has declined over time. The evidence thus far is mixed, but overall suggests that affordability – particularly that of housing – may be compromised for many. Several studies have indicated that walkability often carries a housing cost premium when controlling for other locational factors, though the magnitude of the effect varies (Lucchesi et al. 2021; Gunn et al. 2022). Kim and Kim (2020), for example, found that walkability was significantly associated with higher housing prices in Seoul, Korea, but only in areas with relatively low-cost housing. Gilderbloom, Riggs, and Meares (2015) observed a significant positive association between neighborhood walkability and assessed housing values in Louisville, KY (as well as a negative correlation with housing foreclosures), while Boyle, Barrilleaux, and Scheller (2015) found no significant impact of walkability on the prices of single-family homes or condominiums in Miami, FL. More recently, Christie et al. (2023) concluded that the relationship between walkability and housing affordability in Canada was dependent on city size, with more walkable neighborhoods exhibiting higher home values only in the largest cities.

Rather than focus on home value or sales price, other studies have considered housing affordability in combination with transportation costs (i.e., "location affordability") in the context of local income levels. While evidence suggests that total location affordability is higher in "location efficient" neighborhoods with good walkability and transit accessibility (Renne et al. 2016), very low-income households (≤ 35% of area median income) are unlikely to realize sufficient transportation savings to make up for higher housing costs (Makarewicz et al. 2020; Singer 2021). Additionally, location affordability indices are likely to overestimate the transportation savings expected from moving into more transit-accessible neighborhoods, making the affordability of housing particularly crucial (Smart and Klein 2018).

Despite the somewhat contradictory findings, there is growing evidence that lower-income and other vulnerable groups are being socially and economically excluded from the most walkable urban neighborhoods, particularly in urban regions with competitive housing markets. In an examination of walkable housing in the San Francisco Bay area, for example, Riggs (2016) noted that Black residents were more likely to live in less walkable areas. Interviews with minority residents pointed to a variety of push and pull factors including a lack of affordable housing in areas that were more centrally located *and* walkable, a desire to live closer to friends and relatives who already left the city for the suburbs, and the need for more living space (Riggs 2016). A subsequent comparative analysis of neighborhood walkability and social vulnerability in Charlotte, NC, Pittsburgh, PA, and Portland, OR, revealed significant variations among urban regions with differing social and development patterns (Bereitschaft 2017a). In fast-growing, low-density Charlotte, socially vulnerable groups (i.e., low-income, minority, elderly, low education) were overwhelmingly clustered in a suburban ring north, east, and west of the city center with low walkability and transit accessibility. Accessibility to walkable neighborhoods was more equitable in higher-density, growth-constrained Portland, which boasts a more extensive light rail system and urban growth boundary that regulates low-density suburban expansion. Knight, Weaver, and Jones (2018) examined equity in neighborhood walkability in Buffalo, NY, a shrinking city located in the U.S. "Rust Belt." As in Charlotte, the authors found that low-income and Black households were disproportionately located in areas of the city with poor walkability and relatively low rates of automotive ownership. The more walkable areas of the city were experiencing reinvestment and revitalization, likely resulting in declining affordability (Figure 3.1).

Though scarce, research examining accessibility to walkable environments outside the United States has also begun to uncover a range of place-specific inequities. In Shanghai China, for example, Weng et al. (2019) noted differences in access to "15-minute walkable neighborhoods" (i.e., neighborhoods where most basic goods and services can be accessed within a 15-minute walk), with children and migrants less likely to live within beneficial zones. In an analysis of pedestrian accessibility to neighborhood retail and service establishments in Seoul, Korea, Kim and Park (2020) noted that older residential areas

Figure 3.1 Walkable neighborhoods, such as this one in San Francisco, California, have become unaffordable for many. Credit: Bradley Bereitschaft.

with low property values tended to have poor access to these amenities, though encouragingly they observed a positive correlation between accessibility and concentrations of elderly residents. In Europe, recent work has suggested inequities in access to the more walkable portions of city centers (Bartzokas-Tsiompras, Tampouraki, and Photis 2020) and neighborhoods with superior transit accessibility (Pereira, Vale, and Santana 2023). Several authors have commented on the need for additional research concerning equity in walkability outside of the North American context.

Even among neighborhoods that have similar levels of walkability according to Walk Score® or other macroscale metrics of urban form, actual walking activity may vary due to differences in the safety, design, quality, and maintenance of the streetscape environment. In a walking audit of 76 census tracts in New York City, Neckerman et al. (2009) documented that higher-income areas had significantly more street trees, sidewalk cafés, cleaner streets, and lower rates of crime and vehicular crashes. Bereitschaft (2017b) noted similar differences in the quality of the pedestrian environment in Pittsburgh, PA, with more socially vulnerable (SV) neighborhoods scoring lower in terms of streetscape enclosure, transparency (i.e., presence of windows), complexity, and tidiness. In a nationwide examination of publicly subsidized housing and neighborhood walkability in the United States, Koschinsky and Talen (2015) found that the relationship varied by type of public housing. Importantly, however, among neighborhoods with similar accessibility according to macroscale measures of urban form, the quality of nearby amenities and safety were lower in those with subsidized housing. A recent analysis of pedestrian accessibility and equity in the Colombian cities of Barranquilla and Soledad similarly revealed that lower-income areas were less likely to have paved sidewalks or ramps that improve pedestrian safety (Arellana's et al. 2021). Additionally, the authors found that lower-income areas had significantly fewer amenities and employment opportunities within walking distance relative to wealthier areas. These and other recent studies of inequities in microscale elements of the urban built environment (e.g., Lotfata 2022; Ki and Chen 2023) suggest that even where macroscale measures of walkability indicate sufficient accessibility for disadvantaged groups, deficiencies in the quality of the pedestrian environment and local amenities are not uncommon nor benign.

Advancing Equity in Walkability

In the United States, and in many other regions where automotive-dependent development patterns dominate, enhancing equity in neighborhood walkability will require 1) modifying planning and policy instruments to allow, if not actively encourage, pedestrian oriented development and TOD, 2) ensuring that there is a sufficient stock of affordable housing in walkable communities, and 3) assisting low-income areas realize streetscape improvements and other pedestrian-focused initiatives. As introduced earlier, new urbanism and TOD are two popular "smart growth" strategies that attempt to reduce automotive dependence by increasing density and mixing residential and commercial land use, particularly in close proximity to transit stations. Though these strategies have been implemented to some degree in many cities, increasing density and mixing land use is often at odds with local zoning codes. The adoption of form-based codes (FBCs) is one frequently touted solution. In contrast to conventional zoning codes, which regulate land use type and often needlessly separate commercial and residential uses (as well as different types of residential use), FBCs are used to specify the physical attributes of a building or development, and its relationship with the surrounding area. In addition to allowing mixed-use, FBCs can be used intentionally to create pedestrian-friendly environments by regulating features such as block size, building setbacks and height, façade design, sidewalk design and size, off-street parking, lighting, and balcony types and sizes, among others (Talen 2013a). Together these elements construct much of the "look and feel" of the urban environment, thereby shaping the utility, safety, and overall desirability of walking activity. Despite their potential to help realize more human-centric, walkable urban spaces, FBCs have only been adopted by about 1% of local governments in the United States (Evangelopoulos and Nuworsoo 2016). Additionally, code specifics, and the degree to which they are enforced, can vary considerably from place to place.

Focusing development around transit stations – particularly fixed rail – is another common tactic to curtail urban sprawl and reduce dependence on private transportation. Studies have shown that TOD also has positive impacts on walking behavior, with those who live within TODs and closer to rail stations more likely to walk regularly while driving less (Noland and DiPetrillo 2015; Lu et al. 2018). Not surprisingly, residents of TODs are more likely to walk and use public transit to access work as well as other daily destinations such as grocery stores and schools. A survey of residents living in seven different North American TODs indicated that they were about four times as likely as the average Canadian to achieve weekly recommended levels of physical activity as a result of walking to these and other destinations (Langlois et al. 2016). Where fixed rail systems may not be financially feasible or otherwise practical, bus rapid transit (BRT) systems may be an effective alternative. BRTs are designed to mimic many of the attributes of fixed rail systems resulting in faster and more reliable service than is typical of standard municipal bus systems (Deng and Nelson 2011). As of August 2023, over 180 cities, and some 31 million passengers daily, utilize BRT systems worldwide (BRT+ Centre of Excellence and EMBARQ 2023). While BRT systems may not yet catalyze new development to the same degree as fixed rail systems (Knowles, Ferbrache, and Nikitas 2020), studies have indicated robust positive associations between proximity to BRT and walking activity (Bartels et al. 2016; Lemoine et al. 2016). Whether rail or BRT, local governments often need to make a concerted effort to concentrate on mixed-use development and ensure safe, pedestrian-supportive environments near transit stations to meet TOD goals (Judy 2007).

While instituting smart growth policies including new urbanism and TOD are expected to reduce automotive dependency and shift transportation modes in favor of walking, cycling, and transit, these same policies may also contribute to rising land values and rents that can place significant financial burdens on low-income residents and other vulnerable groups (Chava, Newman, and Tiwari 2018). Additionally, because low- and middle-income households are more likely to utilize public transit, transit systems are expected to see higher ridership and have better long-term viability when supported by mixed-income TODs (Clagett 2014). It is therefore critical that walkable, pedestrian-centric, and transit-oriented places maintain a sufficient stock of affordable housing. Many cities have begun to address this issue by adopting inclusionary zoning/housing policies that either require or encourage private developers to include below-market-rate units in new residential developments. The specifics of inclusionary housing regulations can vary widely, however, with some cities allowing developers to pay into an affordable housing fund or construct affordable housing units off-site in exchange for building market-rate-only developments in more desirable locations (Talen 2013b). Developers have indicated that density bonuses, tax credits, and accelerated permit reviews are likely to be among the most effective incentives (Talen 2013b). In addition to inclusionary zoning, other tools in the affordable housing toolkit include instituting rent control measures, legalizing accessory dwelling units (ADUs), eliminating or reducing parking requirements, establishing affordable housing trust funds, and supporting community land trusts through nonprofit partnerships (Talen 2010; Moos et al. 2018).

Lastly, it is difficult to achieve equity in neighborhood walkability when there exist significant variations in the quality and condition of the streetscape, particularly between low- and high-income areas. Sub-municipal business improvement districts (BIDs) have emerged as a popular mechanism for local business owners to pool resources (typically by levying an additional property tax) and undertake streetscape improvements and maintenance activities such as widening and repairing sidewalks, planting trees, improving street lighting, and providing benches and other pedestrian infrastructure (Wolf 2006). BIDs have also been used to hire security personnel, provide sanitary services, and promote neighborhood amenities and activities (Mitchell 2001). Although these efforts are likely to improve neighborhood walkability and vitality where they are effectively implemented, BIDs have the potential to amplify disparities between high- and low-wealth districts as well as catalyze gentrification (Heller 2017, Elmedni, Christian, and Stone 2018). The capacity to support BIDs and their associated activities is generally lower among smaller commercial areas and lower-income communities. BIDs in wealthier areas of the city may be able to tap into external sources of funding by leveraging political connections and public–private partnerships unavailable to disadvantaged areas (Furman Center for Real Estate & Urban Policy 2007). It is critical then for cities to recognize and perform routine surveys of these inequities so that strategic investments in pedestrian infrastructure can be made, preferably in partnership with BIDs or other community organizations. Additionally, regulations incentivizing or requiring affordable housing, and flexible dues for small businesses, would help mitigate gentrification-induced displacement (Elmedni, Christian, and Stone 2018).

Final Thoughts

Walkability is widely recognized as a key component of healthy, livable, and prosperous urban environments. Yet, the majority of research thus far suggests that walkability is not equally distributed or necessarily accessible to those groups that could benefit most. Advancing equity in neighborhood walkability will require local governments, often in partnership with community groups and with input from residents and businesses, to take an active role in shaping both the macro and microscale urban environment in support of walking, cycling, and transit. Although specific methods and measures will vary from place to place, changes to local zoning codes and housing regulations that favor mixed-use and mixed-income developments in combination with pedestrian-supportive streetscape designs are likely to be broadly efficacious. Particularly where the market has priced out low- and middle-income households, it may be necessary to institute inclusionary zoning and/or other affordable housing policies to ensure equitable access to walkable neighborhoods, as well as promote long-term social and economic resilience. Everyone should have a chance to "live where they love" – to access amenities, services, and opportunities within a safe, human-centric urban environment.

References

Arellana, Julian, Vilma Alvarez, Daniel Oviedo, and Luis A. Guzman. "Walk this way: Pedestrian accessibility and equity in Barranquilla and Soledad, Columbia." *Research in Transportation Economics* 86 (2021): 101024.

Bartels, Clare, Tracy Kolbe-Alexander, Roger Behrens, Sharief Hendricks, and Estelle V. Lambert. "Can the use of bus rapid transit lead to a healthier lifestyle in urban South Africa? The SUN Study." *Journal of Transport & Health* 3, no. 2 (2016): 200–210.

Bartzokas-Tsiompras, Alexandros, Eleftheria Maria Tampouraki, and Yorgos N. Photis. "Is walkability equally distributed among downtowners? Evaluating the pedestrian streetscapes of eight European capitals using a micro-scale audit approach." *International Journal of Transport Development and Integration* 4 (2020): 75–92.

Bereitschaft, Bradley. "Pedestrian exposure to near-roadway PM2.5 in mixed-us urban corridors: A case study of Omaha, Nebraska." *Sustainable Cities and Society* 15 (2015): 64–74.

Bereitschaft, Bradley. "Equity in neighbourhood walkability? A comparative analysis of three large U.S. cities." *Local Environment* 22, no. 7 (2017a): 859–879.

Bereitschaft, Bradley. "Equity in microscale urban design and walkability: A photographic survey of six Pittsburgh streetscapes." *Sustainability* 9 (2017b): 1233.

Bereitschaft, Bradley. "Exploring the spatial intersection of small firm innovation, urban form, and demographics in the Washington, DC, metropolitan area." *The Professional Geographer* (2023) https://doi.org/10.1080/00330124.2023.2215841

Boyle, Austin, Charles Barrilleaux, and Daniel Scheller. "Does walkability influence housing prices?" *Social Science Quarterly* 95, no. 3 (2015): 852–867.

BRT Centre of Excellence and EMBARQ. "Global BRT Data." v. 3.66 (2023). Accessed 15 September 2023: https://brtdata.org/

Brugge, Doug, Allison P. Patton, Alex Bob, Ellin Reisner, Lydia Lowe, Oliver-John M. Bright, John L. Durant, Jim Newman, and Wig Zamore. "Developing community-level policy and practice to reduce traffic-related air pollution exposure." *Environmental Justice* 8, no. 3 (2015): 95–104.

Bueno-Suárez, Carlos, and Daniel Coq-Huelva. "Sustaining what is unsustainable: A review of urban sprawl and urban socio-environmental policies in North America and Western Europe." *Sustainability* 12, no. 11 (2020): 4445.

Burchell, Robert W., and Sahan Mukherji. "Conventional development versus managed growth: The costs of sprawl." *Research and Practice* 93, no. 9 (2003): 1534–1540.

Chava, Jyothi, Peter Newman, and Reena Tiwari. "Gentrification of station areas and its impact on transit ridership." *Case Studies on Transport Policy* 6, no. 1 (2018): 1–10.

Christie, Chelsea D., Christine M. Friedenreich, Jennifer E. Vena, Dany Doiron, and Gavin R. McCormack. "An ecological analysis of walkability and housing affordability in Canada: Moderation by city size and neighborhood property type composition." *PLoS One* 18, no. 5 (2023): e0285397.

Clagett, Tanner M. "If it's no mixed-income, it won't be transit-oriented: Ensuring our future developments are equitable & promote transit." *Transportation Law Journal* 41, no. 1 (2014): 1–32.

Deng, Taotao, and John D. Nelson. "Recent developments in bus rapid transit: A review of the literature." *Transport Reviews* 31, no. 1 (2011): 69–96.

Dumbaugh, Eric, and Wenhao Li. "Designing for the safety of pedestrians, cyclists and motorists in urban environments." *Journal of the American Planning Association* 77, no. 1 (2010): 69–88.

Elmedni, Bakry, Nicole Christian, and Crystal Stone. "Business improvement districts (BIDs): An economic policy or a tool for gentrification." *Cogent Business & Management* 5, no. 1 (2018): 1502241.

Evangelopoulos, Evan, and Cornelius Nuworsoo. "Form-based codes: An overview of the literature," *Focus* 13, no. 1 (2016): 17.

Ewing, R., Amir Hajrasouliha, Kathryn M. Neckerman, Marnie Purciel-Hill, and William Greene. "Streetscape features related to pedestrian activity." *Journal of Planning Education and Research* 36, no. 1 (2016): 5–15.

Ewing, R., and Susan Handy. "Measuring the unmeasurable: Urban design qualities related to walkability." *Journal of Urban Design* 14, no. 1 (2009): 65–84.

Fang, Li, and Fujie Rao. "When industry diversity meets walkability: An analysis of innovation in Baltimore, United States, and Melbourne, Australia." *Journal of Planning Education and Research* 44, no. 2 (2021), 958–969. https://doi.org/10.1177/0739456X211042069

Furman Center for Real Estate & Urban Policy. "The Benefits of Business Improvement Districts: Evidence from New York City." Furman Center Policy Brief, New York University, July 2007. Available at: https://centercityphila.org/uploads/attachments/cit0emsgg001cf6qdxoiz7om6-03-bidnyc.pdf

Gilderbloom, John I., William W. Riggs, and Wesley L. Meares. "Does walkability matter? An examination of walkability's impact on housing values, foreclosures and crime. *Cities* 42, Part A (2015): 13–24.

Glazener, Andrew, and Haneen Khreis. "Transforming our cities: Best practices towards clean air and active transportation." *Current Environmental Health Reports* 6 (2019): 22–37.

Gunn, Lucy D., Tayebeh Saghapour, Billie Giles-Corti, and Gavin Turrell. "Exploring inequities in housing affordability through an analysis of walkability and house prices by neighborhood socioeconomic disadvantage." *Cities & Health* 6, no. 3 (2022): 616–634.

Heller, Erica. "Planning with Improvement Districts." PAS Memo, November-December (2017). Chicago: American Planning Association. Available online: https://www.planning.org/publications/document/9136851/

Herzog, Lawrence. *Global suburbs: Urban sprawl from the Rio Grande to Rio de Janeiro*. New York: Routledge, 2015.

Huang, Yuhan, Chengwang Lei, Chun-Ho Liu, Pascal Perez, Hugh Forehead, Shaofei Kong, John L. Zhou. "A review of strategies for mitigating roadside air pollution in urban street canyons." *Environmental Pollution* 280 (2021): 116971.

India-Aldana, Sandra et al. "Long-term exposure to walkable residential neighborhoods and risk of obesity-related cancer in the New York University Women's Health Study (NYUWHS)." *Environmental Health Perspectives* 131, no. 10 (2023): EHP11538.

Jacobs, Jane. *The death and life of great American cities*. New York: Random House, Inc., 1961.

Johansson, Maria, Catharina Sternudd, and Mattias Karrholm. "Perceived urban design qualities and affective experiences of walking." *Journal of Urban Design* 21, no. 2 (2016): 256–275.

Judy, Meredith H. "The potential for bus rapid transit to promote transit oriented development: An analysis of BRTOD in Ottawa, Brisbane, and Pittsburgh." Master's Thesis. MIT, 2007.

Ki, Donghwan, and Zhenhua Chen. "Walkability inequity in Los Angeles: Uncovering the overlooked role of micro-level features." *Transportation Research Part D: Transport and Environment* 122 (2023): 103888.

Kim, Donghyun, and Jina Park. "Assessing social and spatial equity of neighborhood retail and service access in Seoul, South Korea." *Sustainability* 12, no. 20 (2020): 8537.

Kim, Eun Jung, and Hyunjung Kim. "Neighborhood walkability and housing prices: A correlation study." *Sustainability* 12 (2020): 593.

Knaap, Gerrit, and Emily Talen. "New urbanism and smart growth: A few words from the academy." *International Regional Science Review* 28, no. 2 (2005): 107–118.

Knight, Jason, Russell Weaver, and Paula Jones. "Walkable and resurgent for whom? The uneven geographies of walkability in Buffalo, NY." *Applied Geography* 92 (2018): 1–11.

Knowles, Richard D., Fiona Ferbrache, and Alexandros Nikitas. "Transport's historical, contemporary and future role in shaping urban development: Re-evaluating transit-oriented development." *Cities* 99 (2020): 102607.

Kole, Pieter Jan, Ansje J. Löhr, Frank G. A. J. Van Bellgham, and Ad M.J. Ragas. "Wear and tear of tyres: A stealthy source of microplastics in the environment." *International Journal of Environmental Research and Public Health* 14 (2017): 1265.

Koschinsky, Julia, and Emily Talen. "Affordable housing and walkable neighborhoods: A national urban analysis." *City* 17, no. 2 (2015): 13–56.

Kweon, Byoung-Suk, Jody Rosenblatt-Naderi, Christopher D. Ellis, Woo-Hwa Shin, and Blair H. Danies. "The effects of pedestrian environments on walking behaviors and perception of pedestrian safety." *Sustainability* 13 (2021): 8728.

Lachapelle, Ugo, and Robert B. Noland. "Does the commute mode affect the frequency of walking behavior? The public transit link." *Transport Policy* 21 (2012): 26–36.

Lang, Justin J. et al. "Neighborhood walkability and mortality: Findings from a 15-year follow-up of a nationally representative cohort of Canadian adults in urban areas." *Environment International* 161 (2022): 107141.

Langlois, Myriam, Rania A. Wasfi, Nancy A. Ross, and Ahmed M. El-Geneidy. "Can transit-oriented developments help achieve the recommended weekly level of physical activity?" *Journal of Transport and Health* 3, no. 2 (2016): 181–190.

Lemoine, Pablo D., et al. "TransMilenio, a scalable bus rapid transit system for promoting physical activity." *Journal of Urban Health* 93 (2016): 256–270.

Levasseur, M., Melissa Généreux, Jean-Francois Bruneau, Alain Vanasse, Eric Chabot, Claude Beaulac, and Marie-Michele Bedard. "Importance of proximity to resources, social support, transportation and neighborhood security for mobility and social participation in older adults: Results from a scoping study." *BMC Public Health* 15 (2015): 503.

Leyden, Kevin M. "Social capital and the built environment: The importance of walkable neighborhoods." *American Journal of Public Health* 93, no. 9 (2003): 1546–1551.

Litman, Todd. "Economic value of walkability." *Transportation Research Record* 3, no. 11 (2003): 1828.

Liu, Lu, and Lina Meng. "Patterns of urban sprawl from a global perspective." *Journal of Urban Planning and Development* 146, no. 2 (2020): 04020004.

Lo, Ria Hutabarat. "Walkability: What is it?" *Journal of Urbanism* 2, no. 2 (2009): 145–166.

Lotfata, Aynaz. "Walkable access and walking quality of built environment" In R. Brears (Ed.), *The Palgrave encyclopedia of urban and regional futures*. Springer: Berlin/Heidelberg, Germany (2022): 1–16.

Lu, Yi, Zhonghua Gou, Yang Xiao, Chinmoy Sarkar, and John Zacharias. "Do transit-oriented developments (TODs) and established urban neighborhoods have similar walking levels in Hong Kong?" *International Journal of Environmental Research and Public Health* 15, no. 3 (2018): 555.

Lucchesi, Shanna Trichês, Ana Margarita Larranaga, Helena Beatriz Bettella Cybis, João António de Abreu e Silva, Julian Alberto Arellana, "Are people willing to pay more to live in a walking environment? A multigroup analysis of the impact of walkability on real estate values and their moderation effects in two Global South cities." *Research in Transportation Economics* 86 (2021): 100976.

Makarewicz, Carrie, Prentiss Dantzler, and Arlie Adkins. "Another look at location affordability: Understanding the detailed effects of income and urban form on housing and transportation expenditures." *Housing Policy Debate* 30, no. 6 (2020): 1033–1055.

Makhlouf, Mohamed H. E., Issam Motairek, Zhuo Chen, Khurram Nasir, Salil V. Deo, Sanjay Rajagopalan, and Sadeer G. Al-Kindi. "Neighborhood walkability and cardiovascular risk in the United States." *Current Problems in Cardiology* 48, no. 3 (2023): 101533.

Mehta, Vikas. "Walkable streets: Pedestrian behavior, perceptions and attitudes." *Journal of Urbanism* 1, no. 3 (2008): 217–245.

Mehta, Vikas, and Jennifer K. Bosson. "Third places and the social life of streets." *Environment and Behavior* 42, no. 6 (2010): 779–805.

Mitchell, Jerry. "Business improvement districts and the "new" revitalization of downtown." *Economic Development Quarterly* 15, no. 2 (2001): 115–123.

Moos, Markus, Tara Vinodrai, Nick Revington, and Michael Seasons. "Planning for mixed use: Affordable for whom?" *Journal of the American Planning Association* 84, no. 1 (2018): 7–20.

Nahlik, Matthew J., and Mikhail V. Chester. "Transit-oriented smart growth can reduce life-cycle environmental impacts and household costs in Los Angeles." *Transport Policy* 35 (2014): 21–30.

Nechyba, Thomas J., and Randall P. Walsh. "Urban sprawl." *Journal of Economic Perspectives* 18, no. 4 (2004): 177–200.

Neckerman, Kathryn M., Gina S. Lovasi, Stephen Davies, Marnie Purciel, James Quinn, Eric Feder, Nakita Raghunath, Benjamin Wasserman, and Andrew Rundle. "Disparities in urban neighborhood conditions: Evidence from GIS measures and field observation in New York City." *Journal of Public Health Policy* 30 (2009): S264–S285.

Ng, Wai-Yin, and Chi-Kwan Chau. "A modeling investigation of the impact of street and building configurations on personal air pollutant exposure in isolated deep urban canyons." *Science of the Total Environment* 468–469, no. 15 (2014): 429–448.

Noland, Robert B., and Stephanie DiPetrillo. "Transit-oriented development and the frequency of modal use." *The Journal of Transport and Land Use* 8, no. 2 (2015): 21–44.

Oldenburg, Ray. *The great good place: Cafes, coffee shops, bookstores, bars, hair salons, and other hangouts at the heart of a community*. Washington, DC: Marlowe, 1999.

Park, Yunmi, and Max Garcia. "Pedestrian safety perception and urban street settings." *International Journal of Sustainable Transportation* 14, no. 11 (2020): 860–871.

Pereira, Mauro F., David S. Vale, and Paula Santana. "Is walkability equitably distributed across socio-economic groups? – A spatial analysis for Lisbon metropolitan area". *Journal of Transport Geography* 106 (2023): 103491.

Quistberg, D. Alex, Eric J. Howard, Beth E. Ebel, Anne V. Moudon, Brian E. Saelens, Philip M. Hurvitz, James E. Curtin, and Frederick P. Rivara. "Multilevel models for evaluating the risk of pedestrian-motor vehicle collisions at intersections and mid-blocks". *Accident Analysis & Prevention* 84 (2015): 99–111.

Renne, John L., Tara Tolford, Shima Hamidi, and Reid Ewing. "The cost and affordability paradox of transit-oriented development: A comparison of housing and transportation costs across transit-oriented development, hybrid and transit-adjacent development station typologies." *Housing Policy Debate* 26, no. 4–5 (2016): 819–834.

Riggs, William. "Inclusively walkable: Exploring the equity of walkable housing in the San Francisco Bay area." *Local Environment* 21, no. 5 (2016): 527–554.

Rogers, Shannon H., Kevin H. Gardner, and Cynthia H. Carlson. "Social capital and walkability as social aspects of sustainability." *Sustainability* 5 (2013): 3473–3483.

Singer, Matan E. "How affordable are accessible locations? Neighborhood affordability in U.S. urban areas with intra-urban rail service." *Cities* 116 (2021): 103295.

Smart, Michael J., and Nicholas J. Klein. "*A longitudinal analysis of cars, transit, and employment outcomes*." (2015) San José, CA: Mineta Transportation Institute Publications, College of Business, San José University. Available at: https://scholarworks.sjsu.edu/cgi/viewcontent.cgi?referer=https://www.google.com/&httpsredir=1&article=1198&context=mti_publications

Smart, Michael J., and Nicholas J. Klein. "Complicating the story of location affordability." *Housing Policy Debate* 28, no. 3 (2018): 393–410.

Stoker, Phillip, Andrea Garfinkel-Castro, Meleckidzedeck Khayesi, Wilson Odero, Martin N. Mwangi, Margie Peden, and Reid Ewing. "Pedestrian safety and the built environment: A review of the risk factors." *Journal of Planning Literature* 30, no. 4 (2015): 377–392.

Talen, Emily. "Affordability in new urbanist development: Principle, practice, and strategy." *Journal of Urban Affairs* 32, no. 4 (2010): 489–510.

Talen, Emily. "Zoning for and against sprawl: The case for form-based codes." *Journal of Urban Design* 18, no. 2 (2013a): 175–200.

Talen, Emily. "Prospects for walkable, mixed-income neighborhoods: Insights from U.S. developers." *Journal of Housing and the Built Environment* 28, no. 1 (2013b): 79–94.

Trubka, Roman, Peter Newman, and Darren Bilsborough. "The costs of urban sprawl – Physical activity links to healthcare costs and productivity." *Environment Design Guide* 85 (2010): 1–13.

Urry, John. "Automobility, car culture and weightless travel: A discussion paper." Department of Sociology, Lancaster University, UK (1999). Available online: https://www.lancaster.ac.uk/fass/resources/sociology-online-papers/papers/urry-automobility.pdf

Van den Berg, Pauline, Fariya Sharmeen, and Minou Weijs-Perrée. (2017). "On the subjective quality of social interactions: Influence of neighborhood walkability, social cohesion and mobility choices." *Transportation Research Part A: Policy and Practice* 106 (2017): 309–319.

Weng, Min, Ning Ding, Jing Li, Xianfeng Jin, He Xiao, Zhiming He, and Shiliang Su. "The 15-minute walkable neighborhoods: Measurement, social inequalities and implications for building healthy communities in urban China." *Journal of Transport & Health* 13 (2019): 259–273.

Wolf, James F. "Urban governance and business improvement districts: The Washington, DC BIDs." *International Journal of Public Administration* 29, no. 1–3 (2006): 53–75.

Yeo, Jiho, Sungjin Park, and Kitae Jang. "Effects of urban sprawl and vehicle miles traveled on traffic fatalities." *Traffic Injury Prevention* 16, no. 4 (2015): 397–403.

Zuk, Miriam, Ariel H. Bierbaum, Karen Chapple, Karolina Gorska, and Anastasia Loukaitou-Sideris. "Gentrification, displacement and the role of public investment." *Journal of Planning Literature* 33, on. 1 (2018): 31–44.

4 Near-Roadway Pollution in South Central Fresno, California

Can State-led, Place-based Community Engagement Advance Environmental Justice?

Douglas Houston, Gregg Macey, Jeannine M. Pearce, and Catherine Garoupa

Introduction

Racial and economic disparities in exposure to near-roadway air and noise pollution remain stubbornly persistent due to structural inequalities embedded in the built environment. In 2020, nearly a quarter (24%) of the population of the United States lived near a roadway with a high traffic volume, an increase from about one-fifth (19%) based on 2010 census data (Antonczak et al., 2023). This pattern raises public health concerns given people who live or spend time close to high-traffic roadways are more likely to be exposed to vehicle-related air pollutants such as ultrafine particles which are highly concentrated within 150 meters of roadways (Karner et al., 2010). Exposure to vehicle-related emissions is associated with a broad range of negative health effects including mortality, heart disease, asthma, reduced lung function, and adverse birth outcomes and exposure to roadway traffic noise has been associated with severe sleep disturbance and ischemic heart disease (Boogaard et al., 2022; Hänninen et al., 2014).

These patterns are particularly alarming from an environmental justice perspective given that research has consistently documented racial/ethnic and socioeconomic disparities in traffic exposures nationwide, including heightened exposure to heavy-duty diesel truck traffic, a significant source of roadway noise and pollution (Antonczak et al., 2023; Demetillo et al., 2021). Neighborhoods in the United States with higher concentrations of people of color and lower socioeconomic status are disproportionately impacted by roadway traffic noise (Collins et al. 2020). Schools in the United States with a higher proportion of students of color, higher enrollment, and serving younger children experience greater exposure to roadway noise, which could impact academic performance and student stress and health (Collins et al. 2019). Disparate diesel truck traffic impacts can be particularly intense for residents living along goods movement corridors or near ports or distribution centers (Fried et al., 2024; Houston et al., 2008; Nowlan, 2023).

Such disparities are the result of structural inequalities due to historically uneven land use development, racial segregation, discriminatory lending and

DOI: 10.4324/9781032631110-5

zoning practices, and the exclusion of historically marginalized populations from decision-making (Lane et al., 2022; Rothstein, 2017; Houston et al., 2004). The construction of the Federal Interstate Highway System through low-income areas and communities of color in the center of many metropolitan areas in the United States after World War II divided once vibrant communities, facilitated the relocation of middle- and upper-class residents to the suburbs, and concentrated poverty, transportation infrastructure, and industrial and hazardous land uses in central urban areas, further exacerbating near-roadway environmental justice concerns (Dimento and Ellis, 2013; Rose and Mohl, 2012).

Although federal and state agencies have begun to compile technical guidance on strategies to reduce near-roadway exposures (CARB, 2017), we unfortunately know little about the implementation of these strategies in practice, particularly for historically underserved or disadvantaged communities and the extent to which state-sanctioned, place-based programs can meet dimensions of environmental justice such as participation, distributive fairness, and recognition. We address this gap by reviewing how residents and community leaders in South Central Fresno (SCF), California participated in three place-based initiatives to advance four strategies to address near-roadway impacts. In each, steering committees provided opportunities for residents to actively engage in decision-making processes, priority setting, funding allocation, and program implementation and accountability. Near-roadway strategies included transforming roadways and land use, rerouting trucks from sensitive receptors (i.e., residences, schools, and health care facilities), installing protective vegetative barriers, and limiting freeway and warehouse expansion. We introduce the case study area and initiatives, then recount how residents worked within, across, and outside state-sanctioned, place-based initiatives to prioritize, plan, and implement the four mitigation strategies by engaging land use, transportation, climate, and air pollution across multiple levels of government. We conclude by discussing an important epistemic resource – "multi-level awareness" – that emerged through resident engagement across the initiatives.

South Central Fresno: Place-Based Initiatives to Address Near-Roadway Impacts

The City of Fresno, California's fifth-largest city located in the center of the San Joaquin Valley, provides a valuable case study of how residents have organized through place-based programs to address near-roadway impacts. SCF is one of the most heavily polluted communities in the state due to freight operations, industrial facilities, fossil fuel power generation, and multiple freeways (OEHHA, 2024) (Figure 4.1a). SCF residents experience some of the highest levels of pollution burden and social and health vulnerabilities in California (Werner, 2020).

These stark disparities are the result of a long history of discriminatory policies and disinvestment in South Fresno. Racial segregation was well

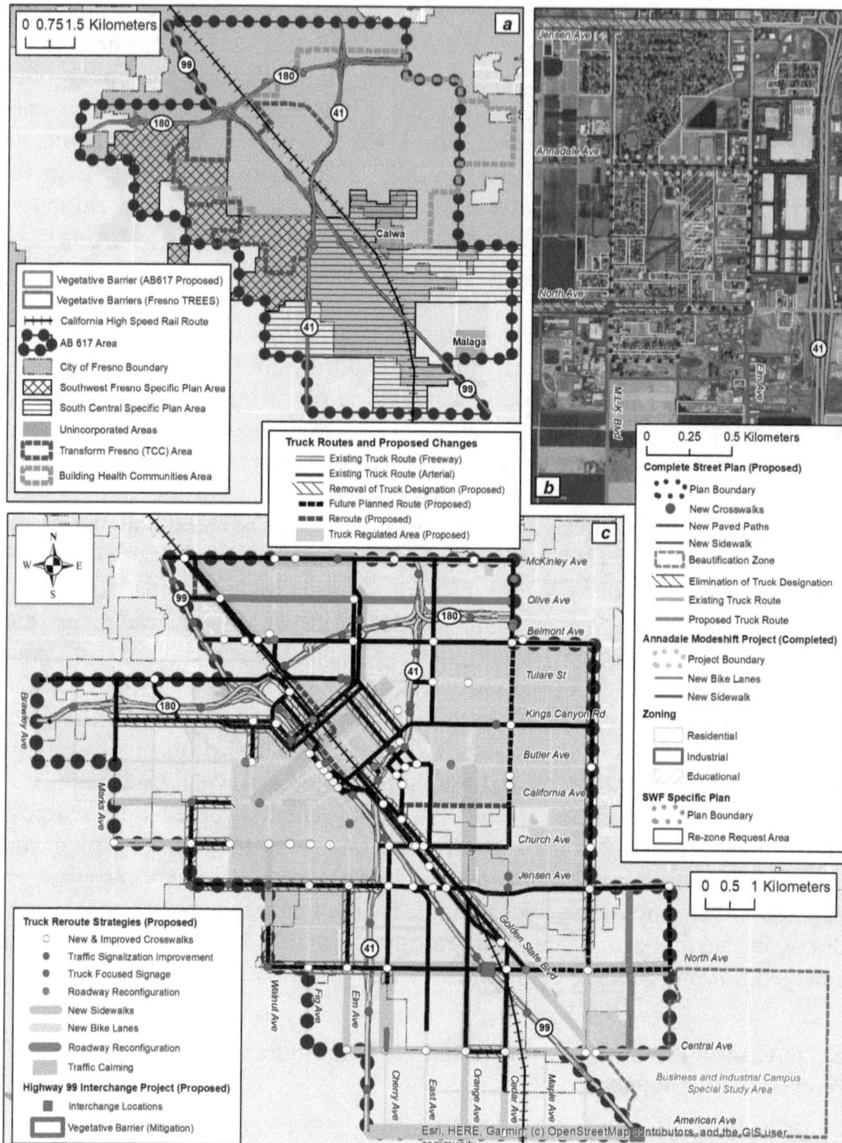

Figure 4.1 South Central Fresno case study area. (a) SCF boundaries (initiative, specific plans, city) and Tree Fresno vegetative barrier locations; (b) Roadway improvements and land use conflicts near North Avenue/Highway 41; (c) Proposed AB 617 roadway and truck reroute plans and Caltrans interchange project locations. Credit: Douglas Houston.

entrenched in the city through racial violence and discriminatory real estate practices by the time it adopted its first general plan in 1918, which further confined industrial operations and lower-income housing to the west and south of downtown (Chacon, 1988). In the 1930s, redlining practices limited the ability of SCF residents to build wealth through home ownership and spurred disinvestment. Urban renewal efforts of the 1950s and redevelopment efforts in the 1960s reinforced these patterns (Zuk, 2013). In the 1950s, Highway 99 was realigned parallel to the railroad tracks, demolishing diverse neighborhoods and further segmenting the city. Highway expansion accelerated white flight and urban sprawl toward northern neighborhoods such that by the year 2000, areas in central Fresno became mixed race or predominately Latino (Zuk, 2013). Although the city's 2002 revised general plan included smart growth principles to revitalize downtown and embrace more compact growth patterns, racial segregation and concentrated poverty were already well established within SCF, which had become a transportation and agricultural hub (Karner et al., 2019).

This chapter examines three place-based initiatives with geographic overlap that used a steering committee approach to engage SCF residents and industry representatives to address a range of environmental, land use, and public health issues, including near-roadway pollution. The first initiative is the City of Fresno's Southwest Fresno Specific Plan (SWFSP), which has been hailed as a model of equitable, community-driven planning (Werner, 2020). Adopted in 2017, SWFSP grew out of resident experiences with the city's 2014 General Plan process. It included guidelines for infrastructure, development, land use, housing, and public health that more directly addressed the needs of SCF residents (City of Fresno, 2017). SCF residents, city staff, community-based organizations, school district representatives, business owners, and developers participated in a public engagement process led by a Project Steering Committee comprised of 21 members who worked closely with city staff and consultants.

The second initiative is Transform Fresno, which is part of the Transformative Climate Communities (TCC) program run by California's Strategic Growth Council (SGC). TCC provides local-level integration in disadvantaged communities for California Climate Investment projects that are funded by the state's Cap-and-Trade program, including renewable energy, low-carbon transport, sustainable land use, and urban greening (Turner et al., 2023). Transform Fresno, which was selected by the SGC and awarded $66.5 million in 2018, grew out of a community engagement process led by a Community Steering Committee (CSC) to produce a shared understanding of community needs and priorities. Residents overwhelmingly approved the plan after an unprecedented level of engagement and the city's largest participatory budgeting effort (Karner et al., 2019).

The third initiative is the Community Air Protection Program administered by the California Air Resources Board (CARB) under Assembly Bill 617 (AB 617). Disadvantaged communities selected under the program engage in monitoring and emissions reduction planning to address neighborhood-scale air quality. The SCF AB 617 initiative engages residents and industry representatives through a CSC that is facilitated by the San Joaquin Valley Air Pollution Control District (SJVAPCD). In 2020, CARB approved the SCF CSC's Community Emissions Reduction Program (CERP), which identifies strategies to reduce emissions of criteria and other air pollutants. As of February 2024, the SCF CERP has been allocated $47 million (SJVAPCD, 2019a; SJVAPCD, 2023).

This case study focuses on four strategies to address near-roadway impacts: transforming roadways and land use, rerouting trucks from sensitive receptors, installing protective vegetative barriers, and limiting freeway and warehouse expansion. Through interviews with AB 617 CSC members (not reported here), we learned that these strategies were deeply informed by participation in the SWFSP and Transform Fresno processes that preceded it, during which residents and community leaders learned about agency processes, requirements, and multilevel governance challenges. Each initiative focused on different but interconnected dimensions of structural inequality in SCF – land use, greenhouse gas (GHG) emissions, transportation, air quality, and public health – that together provide a more comprehensive understanding of the challenges and potential for addressing near-roadway impacts via community-engaged processes.

Our case study is based on an analysis of a variety of sources. We utilized studies of the history, goals, and actions of three initiatives in the peer-reviewed academic literature, media accounts of related projects and decision-making processes, and initiative and agency documents and media. In addition, we reviewed the approved SWFSP and related city council meeting agendas; AB 617 CERP documents, monthly CSC meeting agendas, agency presentations, and meeting recordings; and Caltrans environmental review documents, public comments, and interagency communications. Case study maps were generated using ArcGIS based on data from the City of Fresno's GIS portal and initiative websites. Roadway improvement and land use GIS layers (Figure 4.1b) were adapted from the City of Fresno complete streets plan and the Transform Fresno 2023 Progress Report (City of Fresno, 2015; Turner et al., 2023). GIS layers for the proposed AB 617 roadway and truck reroute plans (Figure 4.1c) were obtained from the AB 617 Community Truck Remote Study (ARCADIS et al., 2024). Finally, the GIS layers for the Caltrans interchange project were adapted from the project's Final Environmental Impact Report (FEIR) (Caltrans, 2023).

Strategies to Address Near-Roadway Impacts

Strategy 1: Transform Roadways and Land Use

The challenges and potential for complete street and city land use strategies to address near-roadway impacts are illustrated by the ongoing zoning disputes in the portion of SCF located west of Highway 41 at North Avenue along the southern city boundary (Figure 4.1b). This focus area contains a mix of residential, educational, and community uses near large distribution centers along Highway 41 to the east and the Cargill meat processing facility to the southwest, facilities which generate heavy-duty truck trips along arterial routes near sensitive receptors.

In 2015, Fresno's "Highway 41 + North Corridor Complete Street Plan" examined truck traffic and transportation infrastructure in this heavily burdened community (City of Fresno, 2015). Funded by the Caltrans Environmental Justice grant program, the plan revealed significant gaps in sidewalks and pedestrian pathways necessary to access nearby schools and highlighted vehicle conflicts and safety concerns. It recommended crosswalk improvements, new signalization and sidewalks along North Avenue, and the installation of paved trails to connect the area's school campuses with residential streets and neighborhood uses nearby (Figure 4.1b). During plan workshops, Cargill facility representatives supported the evaluation of whether truck traffic to their facility should be rerouted to a southern route connecting to Highway 41.

Although the 2015 plan was not implemented, Transform Fresno utilized $343,000 in TCC funding and $150,000 in city funding to complete the Annadale Modeshift Project in the area in 2021. It installed over 14,000 square feet of sidewalk, nearly 1,200 linear feet of Class II bikeways, and signage and street lighting along Annadale Avenue. It closed the gap in pedestrian and bikeway connectivity and enhanced safety for travel among local schools, a community center, a health center, and affordable housing (Turner et al., 2023).

Residents also fought to reduce incompatible, traffic-generating land uses by participating in the development of the SWFSP. Empowered by a public engagement process led by a Project Steering Committee as well as previous experience with the city's 2014 General Plan update, residents succeeded in prohibiting further industrial development in order to promote a more balanced mix of housing, educational, commercial, and park space (Werner, 2020). For example, industrial parcels were rezoned from "Light Industrial" to "Neighborhood Mixed Use," a more pedestrian-supportive urban form that encourages ground-level retail and upper-level housing or office space (City of Fresno, 2017).

The victory was tenuous. While the SWFSP allowed for existing industrial operations to continue as a nonconforming use, property owners filed a rezoning application in April 2021 to revert a 92.5-acre, 15-parcel site back to "Light Industrial," asserting the SWFSP zoning change jeopardized their ability to obtain financing (Figure 4.1b). This action was strongly opposed by residents and organizations who had worked diligently for two years to adopt the SWFSP and by state and regional government officials who said it would enable expansion of industrial operations and undermine state investments through Transform Fresno and AB 617 initiatives to mitigate pollution in one of the state's most heavily burdened environmental justice communities (California Department of Justice, 2022b; CARB, 2021).

Despite the opposition, the Fresno City Council unanimously approved the requested rezone for four of the parcels (36.8 acres) in October 2022 but postponed consideration of the remaining 11 parcels (55.7 acres), directing city staff to seek a compromise between business owners and the community (City of Fresno, 2022). Community representatives and state and regional officials maintain their strong opposition, and the Fresno City Council has yet to act on the rezoning request for the remaining parcels.

Strategy 2: Reroute Trucks away from Sensitive Receptors

To complement its vision to phase out industrial uses, the SWFSP included a community-driven vision for eliminating truck routes that pass through residential neighborhoods to reduce near-roadway air pollution and address safety concerns. Although a SWFSP truck route policy has not been implemented, the SCF AB 617 CSC prioritized and funded a Community Truck Reroute Study for the entire SCF AB 617 area (Figure 4.1c) to update the city's outdated 2005 truck route designations (SJVAPCD, 2019b). The study is managed by the City of Fresno, but a CSC subcommittee was involved in selecting a consulting firm, ARCADIS, to identify, analyze, and evaluate strategies to reduce truck impacts including air and noise pollution, traffic congestion and crashes, bike and pedestrian conflicts, excess wear and tear on roads, and impacts to schools and residential areas while maintaining goods movement operations (ARCADIS et al., 2024).

Interagency coordination was overseen primarily through the study's Technical Steering Committee (TSC) which included city, county, Council of Governments, SJVAPCD, and University of California (UC) Merced staff who provided guidance on logistical and project issues and feasibility. Community input was gathered through the study's Community Advisory Committee, which included SCF AB 617 CSC members, other residents, business leaders, and representatives of community-based organizations and school districts. In addition, the study's community outreach included an online health survey conducted by UC Merced, community-wide pop-up booths and displays, and stakeholder interviews (ARCADIS et al., 2024).

The study's draft recommendations were presented during a January 2024 CSC meeting (SJVAPCD, 2024). They include nine strategies to improve roadway conditions and a map of recommended locations for the improvements and proposed revisions to truck route designations (Figure 4.1c). Recommendations include infrastructure improvements to enhance safety and accessibility for pedestrians and cyclists which align with complete streets and active transportation goals, such as closing gaps in existing sidewalk networks, adding new or restriping crosswalks to increase visibility, and adding dedicated bike travel lanes. Recommendations also include roadway repaving to remove potholes and cracks, traffic calming measures such as speed bumps or narrower lanes to reduce truck speeds, and roadway configurations to alter truck movements and accommodate alternative modes (ARCADIS et al., 2024).

Within the AB 617 planning area, proposed revisions to the existing 2005 truck route network include eliminating truck routes within newly designated "truck regulated areas" that correspond with several areas of SCF where sensitive receptors such as residences and schools are concentrated (Figure 4.1c). This includes the elimination of several routes in the SWFSP area including portions of east-west California, Church, Jensen, and North Avenues and portions of north-south Walnut and Elm Avenues, which could help address conflicts identified in the 2015 complete streets plan. The plan also proposes new alternative truck routes on surface streets along the northwest-southeast Highway 99 and high-speed rail corridor. The unincorporated community of Malaga is designated as a "truck regulated area," even though it currently lacks an adjacent truck route, perhaps in anticipation of proposals to expand industrial uses nearby. The public review of the truck reroute plan was extended through June 2024 and city officials reported to the CSC members that it must be reviewed by the city's planning commission before it is submitted to the city council for approval.

Strategy 3: Install Protective Vegetative Barriers

Residents also devoted considerable attention and analysis to proposals to install roadside vegetative barriers, which consist of continuous plantings of bushes, shrubs, or trees along a freeway or major arterial to limit exposure to particulate matter and gaseous pollutants via dispersion (upward deflection and deceleration) and deposition (e.g., interception and settling). Vegetative barriers offer a tangible, near-term mitigation option for highly impacted locations for which longer-term strategies such as fleet conversion or truck rerouting are more distant, less targeted, or marred by interagency and funding delays. Their effectiveness varies based on the type of near-roadway pollutant, barrier configuration, and plant composition.

Few studies have examined the impact of near-roadway vegetative barriers on gaseous air pollutants even though urban vegetation in general can uptake and permanently remove gaseous air pollutants through absorption into leaf stomata and plant surfaces (Linden et al., 2023). Although one study documented near-roadway trees were associated with a 24–56% reduction in carbon monoxide, the few available measurement studies indicate that vegetative tree barriers are associated with both reduction and increase in near-roadway nitrogen dioxide concentrations (Abhijith and Kumar, 2019).

Vegetative barriers are more consistently associated with reductions in near-roadway concentrations of particulate air pollution. Dense and porous vegetation can mix and dilute polluted air. Particles can also deposit on leaf surfaces upon contact, thereby reducing their overall concentrations in ambient air. Expected reductions vary by the category of particulate air pollution considered. "Ultrafine" particles, or particles less than 100 nm in diameter, typically refer to the smallest exhaust particles in fresh emissions and are measured in terms of particle number concentration (PNC) rather than mass concentration metrics. PNC decreases rapidly downwind of roadways, falling by at least 50% within 150 meters and reaching background levels roughly 400 meters from a roadway (Karner et al., 2010). When vegetative barriers are placed downwind from roadways, they can be associated with a roughly 30% reduction in UFP concentrations within 10 meters of roadways (Abhijith and Kumar, 2019). Deposited particles, however, can be resuspended in ambient air during strong winds, washed from leaf surfaces during rainfall, or deposited on the ground via fallen leaves, resulting in soil or water contamination (Baldauf, 2017).

To maximize the effectiveness of vegetative barriers, they should be constructed as a continuous barrier within meters of high-traffic arterials and freeways that pass near sensitive receptors. Generally, taller barriers result in greater reductions in downwind pollutant concentrations by forcing air up and over vegetation, increasing dispersion. In addition, less porous, more dense vegetation reduces air flow and passage of air pollutants through the barrier. Where possible, multiple rows and types of vegetation such as a combination of bushy shrubs and trees should be installed to provide full vegetative coverage from ground level to the top of the canopy. Trees and shrubbery that do not undergo substantial seasonal change, such as coniferous plants, could have a more consistent impact throughout the year; vegetation with waxy or hairy leaf surfaces tends to remove more particulates than plants with smooth leaves (Baldauf, 2017).

Several projects implemented by Tree Fresno raised awareness in SCF about the potential benefits of vegetative barriers (Figure 4.1a). Since 2016, Tree Fresno has partnered with Sonoma Technology to operate Fresno TREES, a Supplemental Environmental Project (SEP) funded by CARB. SEPs are environmental or public health projects that are designed to provide tangible benefits to communities affected by violations of environmental laws (CARB, 2022). In 2000, Fresno TREES worked with Caltrans to install a row of trees between residences and Highway 99 near Highway 41 (Lopez, 2020). In 2022, Fresno TREES planted a barrier between Tehipite Middle School's athletic fields and the interchange of Highways 41 and 180 in SCF (Panoo, 2022).

The SCF AB 617 CSC also allocated $1 million toward a vegetative barrier strategy and selected Tree Fresno to implement 2–3 vegetative barriers within the AB 617 boundary area. In December 2022 and January 2024 presentations to the CSC, Tree Fresno described their partnership with the City of Fresno to implement the first vegetative barrier within the center median of Golden State Boulevard between Fourth Street and Orange Avenue (SJVAPCD, 2022a; SJVAPCD, 2022b). This northwest-southeast roadway segment runs between and parallel to Highway 99 and the tracks of the future California High Speed Rail project (Figure 4.1a). Although a Tree Fresno representative stressed that this barrier would reduce overall air pollution within this major transportation corridor, residents observed that the proposed barrier location is surrounded by vacant, industrial, and commercial properties. They questioned why the community-approved intervention was not sited near sensitive receptors and understood it would do little to reduce pollution for residents closest to the transportation route, including residents of the unincorporated census-designated place of Calwa adjacent to the high-speed rail tracks and a multitrack railroad transfer yard. Tree Fresno reiterated their experience in implementing vegetative barrier projects and stressed that additional projects could be located adjacent to schools or public housing in the future.

Strategy 4: Limit Freeway and Warehouse Expansion

SCF residents and organizations also organized to address potential increases in near-roadway pollution by opposing freeway and warehouse expansion and the attendant increase in heavy-duty truck traffic through SCF. In January 2023, Caltrans released its FEIR for the South Fresno State Route 99 Corridor Project in South Fresno with a Finding of No Significant Impact on the human environment (Caltrans, 2023). The proposed project seeks to rebuild two existing partial interchanges along Highway 99 where it intersects the east-west routes of North and American Avenues (Figure 4.1c). Constructed in the 1960s, the interchanges do not meet current Caltrans design standards, have pavement that is old and deteriorated, and have tight dimensions that make them difficult to navigate. The project seeks to reconstruct the overcrossings of North and American Avenues to create interchanges with full on- and off-ramps, sidewalks, curbs, lighting, signalization, and drainage improvements.

As part of their project outreach, Caltrans staff made two presentations to the SCF AB 617 CSC. In July 2019 during the project's initial notification phase, resident CSC members asked to be included in a more substantial way throughout the Caltrans planning process and recommended that Caltrans consider vegetative barriers as a mitigative action given Caltrans' previous partnership with Tree Fresno (SJVAPCD, 2021). In December 2021, during the public comment period for the Draft Environmental Impact Report (DEIR), resident CSC members expressed disappointment that Caltrans had not engaged them further, that the CSC had not been directly notified about the DEIR comment period, and that the DEIR was not made available in Spanish until CSC members requested it, leaving many with inadequate time to review its contents (SJVAPCD, 2021).

Additional CSC member concerns align with those outlined in a December 2021 DEIR comment letter by the Leadership Counsel for Justice and Accountability (LCJA), an environmental justice nongovernmental organization based in Fresno. First, residents contested DEIR claims that no sensitive receptors were located close to the project area. They documented residential areas, schools, and places of worship located within two miles of the project, as well as a 1,400-bed juvenile detention center immediately adjacent to the American Avenue interchange. Caltrans responded that the DEIR incorrectly stated that a 2-mile distance was used to identify impacted populations; the FEIR applied a half-mile distance and claimed that there were no sensitive land uses within that radius. Second, residents called for Caltrans to coordinate with the City of Fresno's South Central Specific Plan, under development, which will reconsider the distribution of industrial land uses immediately west of the interchange projects (Figure 4.1c). They also called for Caltrans to delay project approval until completion of the AB 617 Community Truck Reroute Study, which will provide a more comprehensive vision for rerouting heavy-duty trucks. Caltrans rejected the need to align the project with land use and transportation planning efforts for surrounding areas. Third, residents argued that expanding the capacity of two Highway 99 intersections would support the continued expansion of industrial operations within SCF and unincorporated areas nearby. Again, Caltrans stated that they had no legal authority over land use or development decisions and that the interchange project was necessary regardless of future land use decisions. They concluded that the project alone would not cause an increase in traffic, which was projected to grow regardless of the project. In fact, they indicated that without improvements to the deteriorating interchanges, highway congestion would increase and result in deteriorating air quality (SJVAPCD, 2021; Caltrans, 2023).

Concerns that the interchange project was linked to industrial expansion are warranted given the County Board of Supervisors' recent approval of comprehensive revisions to the Fresno County General Plan. The updated General Plan designates 2,940 acres east of the Caltrans projects as a Business and Industrial Campus Special Study Area (Figure 4.1c) (County of Fresno, 2024). Emails between Caltrans and Fresno County staff indicate that Caltrans was aware of and considered this planned industrial expansion in its project design but did not account for increased heavy-duty truck traffic associated with industrial expansion as part of the environmental review (Friends of Calwa and Fresno Building Health Communities, 2023). Furthermore, the California Attorney General claimed that Fresno County's Draft General Plan did not meet state requirements under SB 1000 to "address unique or compounded health risks in disadvantaged communities" because it prioritized large-scale industrial development without taking adequate measures to reduce pollution exposure for Calwa and Malaga (California Department of Justice, 2022a).

In response to community concerns raised to Caltrans by the SCF AB 617 CSC, the FEIR includes a proposal to build a vegetative barrier within SCF as part of the project's GHG mitigation measures (SJVAPCD, 2021). The proposed mitigation includes a sidewalk and adjacent vegetative barrier along the eastern side of South Cherry Avenue from Central Avenue to the Orange Center Elementary School with the stated goal of reducing GHG emissions (Figure 4.1c). The proposed segment is located over 3.5 kilometers east of the intersection of Highway 99 and Central Avenue, which is located midway between the two project interchanges. It is over 300 meters east of Highway 41, a distance at which near-roadway pollutants would likely have already substantially declined. Although this barrier could advance greening goals for SCF and improve aesthetics and pedestrian safety, it is not well placed to reduce pollutant concentrations near a heavily trafficked roadway. Alternatively, a barrier could be placed on the eastern edge of the residential area to reinforce the sparsely vegetated berm separating it from the major Ulta Beauty Distribution Center immediately to the east as well as a major Amazon Distribution Center about 600 meters east along Central Avenue.

After final project approval in January 2023, LCJA and Stanford Law School's Mills Legal Clinic filed a complaint in March 2023 on behalf of Friends of Calwa and Fresno Building Healthy Communities. The complaint raises many of the community concerns that Caltrans failed to address during project review and approval. It asserts that Caltrans and FHWA failed to acknowledge sensitive receptors, consider the cumulative impacts of the project, analyze the project's impacts on vehicle miles traveled under CEQA, or prepare an Environmental Impact Statement for the project under NEPA. In December 2023, the U.S. Secretary of Transportation ordered a reevaluation of the FHWA's Clean Air Act conformity determination for the Highway 99 interchange expansions. In response, agencies and the public will be able to comment on elements of the technical analysis that resulted in the conformity determination (Weaver, 2024).

Discussion

Our case study of state-sanctioned, place-based programs to address environmental impacts featured a variety of approaches to address near-roadway pollution. The three initiatives considered in this chapter operated within and across multiple sectors and policy frameworks, including land use, transportation, climate policy, and air quality. Each sector structures and limits problem identification, strategy development, and public participation in decision-making in unique ways. Working within and across each sector holds the potential to instill "multi-sectoral" awareness, build knowledge of governmental processes, and establish relationships with agency officials.

For example, residents engaged in general plan updates learned how land use planning can solidify a particular vision for future development that can be instilled into legally binding zoning ordinances. This awareness informed robust participation during SWFSP development that resulted in a community-informed vision to phase out industrial land uses. Residents gained an understanding of connections among zoning, traffic generation, and air pollution which led to a truck rerouting proposal, a strategy being implemented on a larger scale through the SCF AB 617 Community Truck Reroute Study. These multisectoral approaches were complemented by an SCF AB 617 CSC-approved vegetative barrier strategy to reduce near-roadway exposures and Transform Fresno's active transportation roadway projects to reduce vehicle travel and GHG emissions. The same organizations that gained multisectoral awareness and knowledge through these processes are now at the forefront of bringing legal action to implement a community-engaged environmental review for the proposed Highway 99 interchange expansion.

This case study also demonstrates the importance of multilevel approaches and oversight in addressing near-roadway impacts. Both the Transform Fresno and AB 617 initiatives resulted from state legislation that directed substantial resources to spur local investment and mitigative actions in disadvantaged communities. These programs enable the California SGC and the CARB to influence local implementation through a competitive process of community selection and approval of community-proposed climate projects and emission reduction strategies, respectively. This structure enhanced multisectoral and multilevel awareness of the substantial interagency coordination necessary to implement regulatory, incentive, and enforcement programs. This case also instilled awareness that higher levels of government can provide critical oversight of local zoning and infrastructure decisions.

Although we found that the state-led, place-based initiatives analyzed were characterized by significant participation, the impact of resident engagement was often muted due to epistemic injustices embedded in steering committee decision-making (Ottinger, 2024). Each initiative used a steering committee structure through which residents and organizations could engage in decision-making, priority setting, funding allocation, and program implementation and accountability (Karner et al., 2019; Nguyen et al., 2022). While multilevel awareness led residents to prioritize four strategies to address near-roadway impacts and push for the multi-sectoral collaboration necessary for implementation, steering committee members were largely relegated to a consultative role in the design of the four near-roadway strategies analyzed. CSC meetings did increase the acknowledgment and consideration of resident concerns derived from their embodied experiences living within impacted communities. However, local knowledge was often deflected or dismissed within state-led processes that were heavily driven by agency priorities and timelines, and which deployed scientific justifications and interagency policy knowledge to deprioritize or obfuscate community concerns, often in favor of industry priorities and voluntary and incentive-based regulatory approaches (Harrison and Contreras, 2023).

Despite these limitations, our case study details several areas in which resident engagement influenced implementation strategies, secured substantial state mitigation funding, and provided multisectoral and multilevel awareness that community groups are now leveraging to advance a broad-based environmental and health justice agenda within the San Joaquin Valley. Future research should consider resident and community leader perspectives on how place-based participatory processes can more effectively utilize an equity lens, support community power sharing, and facilitate transparency, enforcement, and accountable environmental planning.

Acknowledgments

This case study was supported by the UC Institute of Transportation Studies with funding through the State of California Public Transportation Account and the Road Repair and Accountability Act of 2017 (Senate Bill 1).

References

Abhijith, K. V. and Prashant Kumar. "Field Investigations for Evaluating Green Infrastructure Effects on Air Quality in Open-road Conditions." *Atmospheric Environment* 201 (2019): 132–147.

Antonczak, Brittany, Tammy M. Thompson, Mindi W. DePaola, and Gregory Rowangould. "2020 Near-roadway Population Census, Traffic Exposure and Equity in the United States." *Transportation Research Part D: Transport and Environment* 125 (2023): 103965.

ARCADIS, City of Fresno, and San Joaquin Valley Air Pollution Control District. *South Central Fresno AB 617 Community Truck Reroute Study: Truck Routing and Implementation Strategies Report*. Fresno, CA: April 2024. https://www.fresno.gov/publicworks/south-central-truck-re-route-study/

Baldauf, Richard. "Roadside Vegetation Design Characteristics that can Improve Local, Near-road Air Quality." *Transportation Research Part D: Transport and Environment* 52 (2017): 354–361.

Boogaard, H., A. P. Patton, R. W. Atkinson, J. R. Brook, H. H. Chang, D. L. Crouse, J. C. Fussell et al. "Long-term Exposure to Traffic-related Air Pollution and Selected Health Outcomes: A Systematic Review and Meta-analysis." *Environment International* 164 (2022): 107262.

California Air Resources Board (CARB). *Strategies to Reduce Air Pollution Exposures Near High-Volume Roadways: Technical Advisory*. Sacramento, CA: California Air Resources Board. April 2017.

California Air Resources Board (CARB). *Chanell Fletcher to Ron Holt re: Southwest Fresno Rezoning Proposal Comment Letter*, Sacramento, CA: California Air Resources Board. April 6, 2021.

California Air Resources Board (CARB). *Annual Enforcement Report*. Sacramento, CA: California Air Resources Board. June 2022.

California Department of Justice. *Rob Bonta to Steven White re: Comment on Fresno County Draft General Plan*, California Department of Justice: Sacramento, CA: March 29, 2022a.

California Department of Justice. *Rob Bonta to Jennifer Clark re: Comment on Southwest Fresno Rezone Project, Plan Amendment Application No. P20-01665*, Sacramento, CA: California Department of Justice. October 12, 2022b.

California Department of Transportation (Caltrans). *South Fresno State Route 99 Corridor Project Final Environmental Impact Assessment/Environmental Assessment*. Fresno, CA: California Department of Transportation. January 2023.

Chacon, Ramon D. "The Beginning of Racial Segregation: The Chinese in West Fresno and Chinatown's Role as Red Light District, 1870s-1920s." *Southern California Quarterly* 70, no. 4 (1988): 371–398.

City of Fresno. *Highway 41 + North Corridor Complete Streets Plan*. Fresno, CA: City of Fresno. September 2015.

City of Fresno. *Southwest Fresno Specific Plan*. Fresno, CA: City of Fresno. October 26, 2017.

City of Fresno. *City Council Meeting Agenda*. Fresno, CA: City of Fresno.October 13, 2022.

Collins, Timothy W., Sara E. Grineski, and Shawna Nadybal. "Social Disparities in Exposure to Noise at Public Schools in the Contiguous United States." *Environmental Research* 175 (2019): 257–265.

Collins, Timothy W., Shawna Nadybal, and Sara E. Grineski. "Sonic Injustice: Disparate Residential Exposures to Transport Noise from Road and Aviation Sources in the Continental United States." *Journal of Transport Geography* 82 (2020): 102604.

County of Fresno. *Fresno County General Plan*. Fresno, CA: County of Fresno. January 25, 2024.

Demetillo, Mary Angelique G., Colin Harkins, Brian C. McDonald, Philip S. Chodrow, Kang Sun, and Sally E. Pusede. "Space-Based Observational Constraints on NO_2 Air Pollution Inequality from Diesel Traffic in Major US Cities." *Geophysical Research Letters* 48, no. 17 (2021): e2021GL094333.

DiMento, Joseph F., and Cliff Ellis. *Changing Lanes: Visions and Histories of Urban Freeways*. Cambridge, MA: MIT Press, 2013.

Fried, Travis, Anne Goodchild, Michael Browne, and Ivan Sanchez-Diaz. "Seeking Equity and Justice in Urban Freight: Where to Look?" *Transport Reviews* 44, no. 1 (2024): 191–212.

Friends of Calwa Inc. and Fresno Building Healthy Communities v. California Department of Transportation and Federal Highway Administration, *Complaint for Declaratory and Injunctive Relief*, Case no. 1:23-CV- 00353-JLT-EPG (E.D. Cal.), March 8, 2023.

Hänninen, Otto, Anne B. Knol, Matti Jantunen, Tek-Ang Lim, André Conrad, Marianne Rappolder, Paolo Carrer, Anna-Clara Fanetti, Rokho Kim, Jurgen Buekers, Rudi Torfs, Ivano Iavarone, Thomas Classen, Claudia Hornberg, and Odile C. L. Mekel. "Environmental Burden of Disease in Europe: Assessing Nine Risk Factors in Six Countries." *Environmental Health Perspectives* 122, no. 5 (2014): 439–446.

Harrison, Jill Lindsey and Maya Gabriela-Auiler Contreras. "The Subtle Production of Quiescence: Tracing the Neoliberalization of Environmental Justice Policy Implementation." *Environmental Justice* 16, no. 1 (2023): 10–18.

Houston, Douglas, Jun Wu, Paul Ong, and Arthur Winer. "Structural Disparities of Urban Traffic in Southern California: Implications for Vehicle-related Air Pollution Exposure in Minority and High-poverty Neighborhoods." *Journal of Urban Affairs* 26, no. 5 (2004): 565–592.

Houston, Douglas, Margaret Krudysz, and Arthur Winer. "Diesel Truck Traffic in Low-Income and Minority Communities Adjacent to Ports: Environmental Justice Implications of Near-Roadway Land Use Conflicts." *Transportation Research Record* 2067, no. 1 (2008): 38–46.

Karner, Alex A., Douglas S. Eisinger, and Deb A. Niemeier. "Near-roadway Air Quality: Synthesizing the Findings from Real-world Data." *Environmental Science & Technology* 44, no. 14 (2010): 5334–5344.

Karner, Alex, Keith Brower Brown, Richard Marcantonio, and Louis G. Alcorn. "The View from the Top of Arnstein's Ladder: Participatory Budgeting and the Promise of Community Control." *Journal of the American Planning Association* 85, no. 3 (2019): 236–254.

Lane, Haley M., Rachel Morello-Frosch, Julian D. Marshall, and Joshua S. Apte. "Historical Redlining is Associated with Present-Day Air Pollution Disparities in US Cities." *Environmental Science & Technology Letters* 9, no. 4 (2022): 345–350.

Linden, Jenny, Malin Gustafsson, Johan Uddling, Ågot Watne, and Håkan Pleijel. "Air Pollution Removal through Deposition on Urban Vegetation: The Importance of Vegetation Characteristics." *Urban Forestry & Urban Greening* 81 (2023): 127843.

Lopez, Frank. "Making Like a Tree and Leaving Pollution Behind." *The Business Journal*, April 17, 2020. https://thebusinessjournal.com/making-like-a-tree-and-leaving-pollution-behind/

Nguyen, Peter, Henry Borison, Celine Busnelli, Jessica Raeli Ison, and Jonathan London. *The Past, Present and Future of AB 617; Envisioning a Way Forward Together: Convening Report*. Davis, CA: University of California, Davis. January 2022.

Nowlan, Aileen. *Making the invisible visible: Shining a light on warehouse truck air pollution*. Environmental Defense Fund. (2023): https://globalcleanair.org/files/2023/04/EDF-Proximity-Mapping-2023.pdf

Office of Environmental Health Hazard Assessment (OEHHA). "CalEnviroScreen 4.0." Last modified February 26, 2024. https://oehha.ca.gov/calenviroscreen/report/calenviroscreen-40

Ottinger, Gwen. "Careful Knowing as an Aspect of Environmental Justice." *Environmental Politics* 33, no. 2 (2024): 199–218.

Panoo, Ashleigh. "Can Trees Reduce Pollution at Schools Next to Freeways? A Fresno Campus Tries Plantings." *fresnoland*, December 2, 2022. https://fresnoland.org/2022/12/02/can-trees-reduce-pollution-at-schools-next-to-freeways-a-fresno-campus-tries-plantings/

Rose, Mark H., and Raymond A. Mohl. *Interstate: Highway Politics and Policy Since 1939*. Knoxville, TN: Univ. of Tennessee Press, 2012.

Rothstein, R. *The Color of Law*; Liveright Publishing Corp.: New York, 2017.

San Joaquin Valley Air Pollution Control District (SJVAPCD). *South Central Fresno AB 617 Community Steering Committee Meeting*. Fresno, CA: San Joaquin Valley Air Pollution Control District. July 10, 2019a.

San Joaquin Valley Air Pollution Control District (SJVAPCD). *Community Emissions Reduction Program: South Central Fresno*. Fresno, CA: San Joaquin Valley Air Pollution Control District. September 19, 2019b.

San Joaquin Valley Air Pollution Control District (SJVAPCD). *South Central Fresno AB 617 Community Steering Committee Meeting*. Fresno, CA: San Joaquin Valley Air Pollution Control District. December 14, 2021.

San Joaquin Valley Air Pollution Control District (SJVAPCD). *South Central Fresno AB 617 Community Steering Committee Meeting*. Fresno, CA: San Joaquin Valley Air Pollution Control District. December 13, 2022a.

San Joaquin Valley Air Pollution Control District (SJVAPCD). *Urban Greening and Vegetative Barriers Proposal Presentation to South Central Fresno AB 617 Community Steering Committee*. Fresno, CA: San Joaquin Valley Air Pollution Control District. December 13, 2022b.

San Joaquin Valley Air Pollution Control District (SJVAPCD). *Community Emissions Reduction Program: South Central Fresno 2023 Annual Report*. Fresno, CA: San Joaquin Valley Air Pollution Control District. December 14, 2023.

San Joaquin Valley Air Pollution Control District (SJVAPCD). *South Central Fresno AB 617 Community Steering Committee Meeting*. Fresno, CA: San Joaquin Valley Air Pollution Control District. January 17, 2024.

Turner, V. Kelly, Jason Karpman, Elena Hernandez, Samantha Astudillo, Lauren Dunlap, and Bo Liu. *Transform Fresno: 2023 Progress Report on Implementation of the Transformative Climate Communities Program Grant*. UCLA Luskin Center for Innovation, Los Angeles, CA: May 1, 2023.

Weaver, Gregory. "Biden Administration Steps in on Caltrans' Highway Expansion in Fresno". *Streetblog*, January 12, 2024.

Werner, Ashley. *Southwest Specific Plan, City of Fresno: Community-Driven Planning for Equitable & Healthy Neighborhoods*. Sacramento, CA: Governor's Office of Planning and Research. 2020.

Zuk, Miriam Zofith. "Health Equity in a New Urbanist Environment: Land Use Planning and Community Capacity Building in Fresno, CA." PhD diss., University of California, Berkeley, 2013.

5 Understanding Environmental and Food (In)justice through Community-based Research

Stories from Sacramento Community Coalitions

Leah Joyner, Gabrielle Gonzales, Si Gao, and Adrienne Cachelin

Introduction: Food Access, Justice, and Sovereignty

Despite Sacramento's recognition as "America's Farm-to-Fork Capital," food insecurity, or a lack of access to nutritious, healthy, and culturally appropriate food, is a persistent issue across the region. According to the 2021 Sacramento Region Food System Action Plan (SRFSAP), nearly 300,000 residents (one in six) across the four-county region experience food insecurity (Valley Vision, 2021).

Three paradigms have been suggested to describe approaches to food insecurity including food security, food justice, and food sovereignty (Trauger, 2017). The food security paradigm, rooted in a neoliberal framework, promotes food access and hunger alleviation programs such as food banks, which many scholars argue is a Band-Aid approach that skirts the root issues (Fisher, 2017). As a more progressive approach, the food justice paradigm promotes alternatives to the corporate food system such as sustainable agriculture and local food purchasing. The food sovereignty paradigm advocates dismantling the corporate food system in favor of autonomous community control and democratization of the food system (Holt-Giménez & Shattuck, 2011; Trauger, 2017). Food sovereignty has been defined as, "the right of peoples to healthy and culturally appropriate food produced through ecologically sound and sustainable methods, and their right to define their own food and agriculture systems" (Nyéléni, 2007, p. 7). Previous research indicates that in urban areas, including Sacramento, Black, Indigenous, and People of Color (BIPOC) frequently report the highest levels of interest in urban agriculture, yet due to systemic barriers, experience the lowest levels of access to community gardens (Joyner et al., 2022; Reese, 2019; Valley Vision, 2021).

The political and economic systems that produce racially inequitable access to food and urban agriculture shape what is often referred to as food apartheid or the systemic production of food inequity through intentional racial and economic policies and practices that produce racial and class-based

DOI: 10.4324/9781032631110-6

inequalities in community food access (Washington, 2018, as cited in Brones, 2018; Reese, 2019). The systems that produce uneven food access are intertwined with inequalities in environmental health, educational access, and healthcare (IPESFS, 2020; Bentley, 2014). The food sovereignty paradigm requires that strategies to combat these systemic inequalities must be rooted in community autonomy and self-determination.

The food sovereignty movement has been championed by La Via Campesina (LVC), an influential transnational movement comprised of peasant and family farmers unified against the corporatization of the food system and neoliberal policies that threaten agrarianism (Martínez-Torres & Rosset, 2010; Peña, Calvo, McFarland, Pancho, & Valle, 2017). Central to LVC's practice of food sovereignty is agroecological education, which employs "Freirean political philosophies of education to influence agrarian practices in ways that incorporate understandings of landscape histories, public policies, and cultural traditions" (Mann, 2019). Thus, as a critical component of food sovereignty, agroecological education has the potential to prompt powerful paradigm shifts leveraging culturally specific food knowledge and empowering individuals to interrogate and disrupt political processes that shape food systems, and exploration of the linkages between cultural politics to environmental sustainability (Cachelin et al., 2019; Meek, 2016). For this reason, community-based research is foundational to uncovering pathways for food sovereignty.

Stories from Sacramento

Because community-engaged research is foundational to food sovereignty, researchers began with exploratory interviews with community organizations working on food justice, environmental and educational equity, and urban agriculture. In Sacramento, we employed an interview methodology adapted from the semi-structured interview guide designed to promote reflexivity on the part of both the researchers and interviewees (Joyner, Yagüe, & Cachelin, 2023). Interviewees included representatives of local urban agriculture operations, nonprofits, and community coalitions working on issues of food and environmental justice. Contacts were identified via online review of Sacramento-based organizations' mission statements resulting in a list of 25 contacts whom we reached out to via email to arrange for a one-hour interview either via Zoom or in person. Interviewees were offered a $50 gift card in recognition of their time and expertise for participation. Researchers obtained verbal consent to record interviews which were transcribed and analyzed via the qualitative analysis software NVIVO. The interviews were coded inductively using several rounds of coding by two researchers and discussed in collaborative meetings. Through preliminary analysis, we identified three themes that were most frequently discussed by interviewees, and which align with a food sovereignty paradigm: (1) Political Economy and Environmental Justice, (2) Environmental Justice Concerns, and (3) Agroecological Educational Opportunities.

Political Economy and Environmental Injustice

The systemic production of urban food inequity, known as food apartheid, is rooted in institutionalized practices of racism and classism (Washington, 2018, as cited in Brones, 2018; Reese, 2019). The food apartheid framework asserts that "inequity is not a natural occurrence based in ecological limits, but rather an explicit outcome of political economy based in structural racism and unequal geographies of access" (Joyner et al., 2022, p. 68). These unequal geographies of access can manifest through infrastructural inequalities such as racially inequitable access to transportation, funding for local schools, maintenance of sidewalks, water quality and access, green space allocation, and pollution mitigation.

Throughout our interviews, community members pointed to numerous instances of structural and institutionalized racism embedded in local infrastructure that complicate their ability to engage in urban agriculture. For example, the city is currently building low-income housing in a historically underserved neighborhood in Sacramento. A community garden was built adjacent to the development which informed one of our respondents who works in this community garden to share their insights into the happenings of the neighborhood;

> I just talked to someone who moved in here. Their electricity was wrong. So, all their appliances work in the kitchen on the same circuit. So, it was tripping their circuit breaker. They couldn't use their restroom; they had to use the unit next door, and these are brand new units. and I wish I could say I was surprised.

This quote indicates discrepancies in housing insecurity related to socioeconomic status and race, as most residents living in this low-income housing are BIPOC. They went on to say,

> It's also a huge transportation desert. Most of the bus lines have been cut out here. They have a bike program, but from what I hear it's almost impossible for people to get the orientation.

Residents in poorly resourced neighborhoods often face a dearth of transportation options to take them to critical places like work or the grocery store (Tehrani et al. 2019). A respondent who has lived in Sacramento for many years and has been dedicated to working toward environmental justice describes a sense of abandonment of specific areas of Sacramento, specifically those areas with a large population of BIPOC and low-income people, stating,

> If you go out to the heights today, this is what we call del Paso Heights. Some of it still looks like it did in 1940, I mean, truly, no sidewalks, ramshackle houses, and just ramshackle now because of age.

This observation is connected to research that indicates failing infrastructure and neighborhood disinvestment in urban centers across the United States stems from a complex web of housing segregation, historic redlining, and ongoing systemic racism (Reese, 2019; Sbicca, 2012). Similarly, another concern expressed was the accessibility of space for urban farming. One participant noted,

> Availability of space is huge, depending on where people are living. They may or may not have any space that they can grow in, and things like community gardens that could be accessed aren't always available in their neighborhood.

From the abandonment of poor racially diverse sections of the city to a lack of transportation and land, the above quotes demonstrate community members' knowledge of the complex landscape of racial inequality in Sacramento. Another respondent noted an example of racial inequality related to the education system, saying,

> I'll watch a group of refugees land into a neighborhood, and they're all sent on to the same school. It wouldn't be a bad thing if that school wasn't already overtaxed and under resourced. They're getting fewer resources. They're already dealing with so many issues coming into that site.

The clustering of low-income students of color, including refugees, in schools with fewer resources has also been tied to historical racial segregation and redlining of neighborhoods, leading to the concentration of poorly resourced public schools in low-income neighborhoods (Palik & Østby, 2023). The collection of racially inequitable infrastructure inequalities cited by our interviewees indicates how food apartheid takes shape in Sacramento and is connected to educational equity, community infrastructure, and the natural environment.

Environmental Justice Concerns

When prompted to elaborate on environmental justice concerns related to urban agriculture in Sacramento, aside from infrastructural inequalities and food access itself, many people cited concerns relating to three main categories: (1) water quality and access, (2) environmental pollutants, and (3) climate change concerns. All of these issues can be tied to urban agriculture.

Water Quality and Access

Water quality and access are necessary components of food cultivation that are often particularly challenging for growers in historically disinvested urban areas with potential (Gregory et al. 2015). Reflecting on the difficulty of establishing community gardens in urban spaces, one grower said,

> Water was one of the main [concerns], and so the lot was an optimal space, but it didn't have water. The city was gonna pay the cost upfront. And then because this is not meant to be a farm or open space like this forever, and whoever ends up buying and developing will repay that cost to the city for already adding water hooked up. So right now, we're just on borrowed time, but I am fighting like hell to keep this.

Other community members echoed this sentiment, identifying water access as a crucial barrier to urban agriculture especially in marginalized neighborhoods. Like many other urban areas, Sacramento has industrial zones and parks where manufacturing and other industrial activities take place. Pollutants may be directly introduced into nearby water bodies via industrial discharge or indirectly introduced by the utility infrastructure and maintenance. These discharges, often laden with heavy metals and toxins, pose a significant risk when not treated appropriately (Delta Independence Science Board, 2018; Laurin & Luengen, 2019).

Sacramento lies within the fertile agricultural region of the Central Valley of California, where numerous dairy farms in the surrounding areas may contribute to nonpoint source pollution. Excessive fertilizers used on crop fields and animal wastes at feedlots can be released into nearby water bodies through agricultural runoff or soil leaching, affecting urban agriculture areas downstream (Hanak et al. 2019). Urban runoff, stemming from residential areas, parking lots, and roads in the metropolitan area, can also contribute to water pollution by transporting sediments, nutrients, heavy metals, and pesticides into nearby water bodies (Yang et al. 2022). These pollutants not only compromise water quality but also threaten the biodiversity of urban agroecosystems and the viability of urban farming. In Sacramento, this is also an issue of racial inequity as the city is among a list of top metro areas in the United States where formerly redlined neighborhoods, which remain primarily inhabited by residents of color, face significantly greater flood risks than more affluent and primarily white non-redlined neighborhoods (Katz, 2021).

Environmental Pollutants

The presence of environmental pollutants is a major concern for community members. One source of pollution mentioned was the Sacramento Executive Airport with one interviewee saying,

> [You] need the clean ground to grow your food, and Meadowview is one of the most highly polluted areas in Sacramento [and] we have the executive airport spewing out the lead. Yes, from the planes. It just keeps piling up.

Indeed, in a 2008 report, the federal Environmental Protection Agency estimated that small aircraft taking off and landing at Sacramento Executive Airport emit 278 kilograms – or 0.3 tons – of lead per year, placing it among the top 6% of airports in the nation (Lillis, 2017). Soil pollution driven by the accumulation of pesticides, herbicides, and heavy metals in soils as a part of historical land use can be a significant barrier to urban agriculture and food security in urban areas. Toxic elements such as Zn, Pb, Cu, and As can also be accumulated in the soil through the use of chemicals such as fungicides or pesticides for disease prevention and treatment, resulting in their accumulation in agricultural products and entering the food chain (Havugimana et al. 2017). Depending on the soil type, microbial activity, and the form of the pollutant (e.g., organic or inorganic, solid or liquid), the residence time of these pollutants in soils can range from decades to centuries and are often found to reduce soil fauna abundance and soil microbial diversity significantly (Chien et al. 2008). Exposure or consumption of these pollutants increases the risk of chronic health issues, including cancer, neurological disorders, reproductive problems, cognitive deficits, and kidney damage (Oliver & Gregory, 2015). These contaminants undermine soil health and human health, necessitating remediation efforts (Khan et al., 2023).

Another environmental issue is exposure to high-tension power lines, which interviewees are concerned about posing various risks to people, soils, and pollinators. One interviewee said,

> ... a lot of folks don't realize ... they're growing under high tension lines. ... microwaves that are coming off those high-tension wires above your head, and the gamma rays that come from that [are dangerous].... So, the environmental justice aspect of it is even though these are open plots, they aren't in the pristine areas of land to grow without the above ground pollution.

This observation is connected to previous research which indicates that farmworkers laboring near these lines face electrical hazards, while construction and maintenance activities may induce soil disturbance and compaction, reducing soil health and water infiltration (Shaheb et al. 2021). Studies have also shown that the aboveground pollinator abundance and biodiversity may also suffer from electromagnetic fields, where honeybees' cognitive and motor abilities are found to be significantly impaired under low-frequency electromagnetic fields (Shepherd et al. 2018). Electromagnetic fields may also affect plant health, although this aspect warrants further investigation (Goldsworthy, 2006).

Climate Change Concerns

Community members also identified the impacts of climate change as critical. The impacts of climate change, including deteriorating air quality, water scarcity, intensified heatwaves, and flooding events pose well-documented barriers to urban agriculture. For instance, rising temperatures and heat waves can stress plants, leading to reduced crop yields and decreased overall productivity. Heatwaves can be particularly damaging, causing wilting, sunburn, and reduced fruit set in plants (Pathak et al. 2018). One interviewee noted the Heat Island Effect, or when "urbanized areas experience higher temperatures than outlying areas" (EPA 2023), stating,

> One of the things I've become really aware of is the Heat Island effect. A lot of families [we work with] are living in low-income neighborhoods which tend to have fewer trees and more asphalt. It really does relate to their ability to grow and or be outside. Some of our students have done some temperature checks and found that their school sites can be upwards of 5 degrees hotter than the East Sacramento neighborhoods that have all the trees.

Participants also described strategies for climate change adaptation in urban agriculture, saying

> This is our response to climate change. We're finding that the grow tunnels actually [need] shade cloth in the summer, and [that] they help reduce temperatures on plants by a significant amount, as well as retain moisture. And then in winter we can close them up and use them like a greenhouse.

Another strategy utilized by a participant includes seed saving from plants to preserve the ability to grow in the future, particularly heirloom crops. They remarked,

> Over 60% of the world's seed is owned by 6 companies. So, understanding, ecologically appropriate ways of saving seed and growing seed that's appropriate for your climate, and especially with climate change [is] such a major part of everything.

Additionally, prolonged droughts from climate change can lead to water shortages for irrigation, while excess water caused by intense rainfall events and increased flooding in some areas can oversaturate soil. This process will induce anaerobic conditions that hinder soil microbial respiration, as well as accelerate the emissions of nitrous oxide (a greenhouse gas that has a 273 times stronger global warming potential than CO_2 over a 100-year period) via incomplete denitrification (Livesley et al. 2010). Eventually, flooding events will likely drive topsoil erosion, causing significant damage to urban farms (Borrelli et al. 2020). Air pollutants such as particulate matter, nitric acids, and photochemical smog during extended long summers in Sacramento can further harm plant and soil microbial life by reducing plant photosynthetic rate and microbial activity and functional diversity (Manisalidis et al. 2020). These environmental challenges disproportionately affect disadvantaged communities and exacerbate health issues in populations already vulnerable to such disparities.

Many interviewees located their concerns about climate change within the larger scope of their visions for the future of the food system. When asked to describe their vision for an ideal food system in Sacramento, several community members pointed to the need for a complete overhaul of the agricultural system, with one person saying,

> The system of farming that has happened historically is out of date, like we've been just destroying topsoil since World War 2 and using synthetic fertilizers that whole time and pesticides and stuff. So, there's a need to create a new generation of farmers who can do agriculture in an ecological way.

This quote indicates that the type of community-based agroecological education that Growing Educational Pathways for Food Sovereignty (GEPFS) promotes may align with community members' hopes and dreams to combat climate change through urban agriculture.

Agroecological Educational Opportunities

In addition to questions related to environmental justice concerns, we also asked interviewees if they felt an educational program designed to promote food sovereignty might be needed in Sacramento. Many respondents reported that such a program could be impactful in Sacramento for students from marginalized communities. Several interviewees provided additional insights into the role of agroecological education in the pursuit of food sovereignty with the founder of an urban garden in Sacramento stating,

> That's the best part about it for me – when we have the youngsters come out here and they get inspired, and they run out, and they pick the vegetables themselves, and they eat that stuff because not often are very many places for them to do that. So that's what gives me joy. They're like, Wow! And I think they gain respect for the land, and we talk about it. We hope that that translates over into other things that they do.

Another interviewee who facilitates urban agriculture educational initiatives at a local nonprofit considered how their programs not only teach students about how to grow food but also empower young people to question the roots of food insecurity which cultivates empowerment, another key component of a food sovereignty paradigm;

> They don't need us really… they just need us to help, guide, or support, [to] pique their interest and develop their curiosity… it also is [about] empowering people to have the voice to call out that the situation in their community is not okay… and so building their awareness, and I think empowers them to have a voice.

Interviewees' perspectives on urban agriculture as a tool for empowerment align with the aims of agroecological education that seeks to foster deep relationships with the environment not just by growing food but also by interrogating the linkages between sociopolitical structures and the natural environment (Meeks, 2016). It is this approach to disrupting food apartheid via food sovereignty-oriented community-based agroecological education that guides the GEPFS model.

Implementing GEPFS

In this chapter, we have explored our initial findings from interviews with community members working in the field of urban agriculture and food justice in Sacramento regarding their perspectives on localized intersections between food equity, urban agriculture, and environmental justice. In presenting direct quotes from community members, we tried to let the data speak for itself as a contextualization of community visions for food sovereignty in Sacramento. The application of an environmental justice framework clarifies the ways that political and economic factors connect to community members' views on food sovereignty.

In thinking through how the agricultural community can work with academics to foster food sovereignty, we considered how the GEPFS project in Salt Lake City (Cachelin et al., in press) might be adapted to Sacramento. GEPFS emerged in response to the complex relationships between food sovereignty, political economic discourse, and educational inequities as a community-based agroecological educational model designed to promote food sovereignty and educational equity through urban agriculture-based pathways to higher education. GEPFS was designed by a collective of resident leaders, organizers, educators, and researchers with the SPARC Environmental Justice Lab at the University of Utah, unified by the idea that culturally sustaining urban agriculture, when connected to schools and universities, can increase the ability of marginalized students to assert their right to belong in schools and foster local autonomous food production (Cachelin et al., in press). GEPFS is both an applied program that establishes garden-based educational opportunities with current and future university students, and a research project that contributes to the advanced theoretical understanding of how food justice education can promote the right to belong among culturally and linguistically diverse (CLD) students (Kuttner, 2023). GEPFS builds on existing research establishing that community garden programs, culturally responsive/sustaining pedagogies, and family engagement can promote food justice and engender student belonging (Borck, 2020; Guitart et al., 2012; Osterman, 2000).

Our research suggests that GEPFS may also be a good fit for Sacramento, given that low-income and communities of color in Sacramento report the highest levels of interest and lowest levels of access to community gardens (Valley Vision, 2021). In addition, food insecurity is prevalent among students at the local university, California State University, Sacramento (CSUS), as just over 47% report experiencing food insecurity (Chancellors Office Study on

Basic Needs, 2018). The SRFSAP identifies a need for programs supporting food literacy and urban agriculture in K-12, promoting increased access to community gardens and local foods for communities of color and lower-income residents (Valley Vision, 2021).

A dual-site approach is designed to explore the potential of replicating and adapting GEPFS in unique community contexts. The core components of the model are as follows.

The components of the GEPFS model outlined in Figure 5.1 demonstrate how community-based research can both uncover and disrupt political-economic factors that shape localized manifestations of food apartheid while also informing the collaborative design of culturally responsive food sovereignty educational programs. This chapter provides an in-depth look at our preliminary research in Sacramento, which involves research into the local political economy, community concerns, and environmental conditions that illuminate how food apartheid takes shape locally while also highlighting opportunities for research partnerships that support community visions for food sovereignty.

Following the model presented in Figure 5.1, our next steps will be to continue to engage in relationship-building fieldwork. This type of longitudinal approach to community connections is critical to the aims of forging close partnerships and positioning researchers to work with community members to co-construct agroecological educational programs.

As we adapt the GEPFS Curriculum to Sacramento, we are considering the unique environmental issues and community considerations highlighted in this

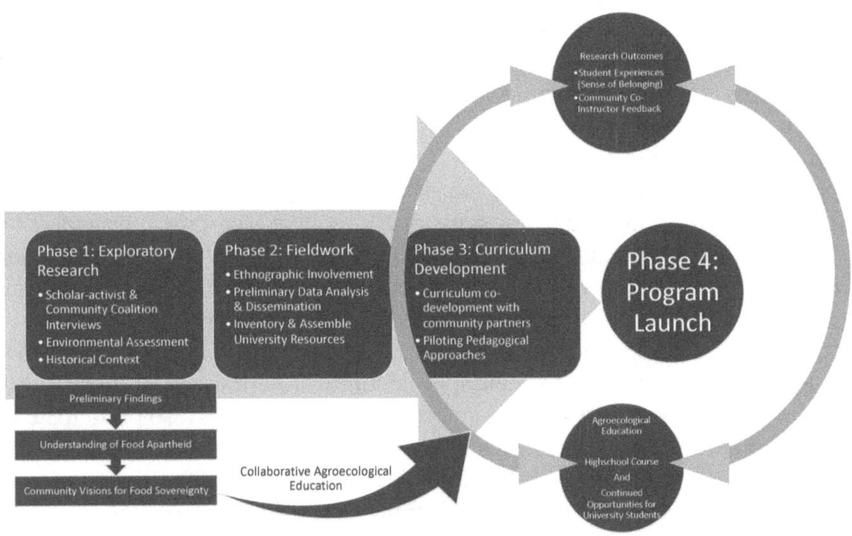

Figure 5.1 The Growing Educational Pathways for Food Sovereignty model. Credit: Authors.

chapter. Based on these findings, we may include several specific urban agriculture techniques that may be relevant to include in the program such as phytoremediation, cover-cropping, drip irrigation, rainwater harvesting, and other methods to prevent soil erosion and mitigate pollution (Gao et al. 2017).

Phytoremediation is a particularly viable solution to address heavy metal and pesticide soil pollution driven by historical land uses (Alaboudi et al. 2018). This technique involves using specific plants to remove heavy metal pollutants from soil, aiding in the restoration of contaminated urban agriculture sites (Awa & Hadibarata, 2020). Sustainable agricultural practices at the ecosystem level are also crucial for addressing the environmental problems in urban agriculture. Techniques such as cover cropping, drip irrigation, and rainwater harvesting can improve water use efficiency and reduce water-related environmental and social barriers. Adding drought-resistant crops and native vegetation not only helps conserve water but also attracts pollinators and acts as a windbreak to mitigate soil erosion (Lin et al. 2015). Other system-level practices include redesigning the agroecosystem. For instance, adding tree canopy cover such as incorporating fruit trees in urban farms can introduce shade to combat the effects of heatwaves while serving as natural air purifiers. Natural features like rain gardens can also be added to small urban farms to buffer stormwater, capture and filter pollutants, and decrease surface water runoff. The following paragraph provides suggestions for how the above environmental information can be incorporated into an education program for high school and college students.

GEPFS is designed for high school and college students to inspire environmental stewardship, promote sustainable urban agriculture, and cultivate an understanding of culturally diverse perspectives on food sovereignty. Through the program, students engage in a variety of activities that teach urban agriculture skills and culinary skills that contribute to a sense of ownership and control in the local food systems. Through GEPFS, students engage in field experiences at several local grassroots mutual aid gardens, worker-run agricultural co-ops, herbal and medicinal collectives, community gardens, and at the University of Utah's Edible Campus Gardens. To honor the diversity of perspectives and unique knowledge that lives in the community and to support intergenerational knowledge transmission, the course is co-taught by a team of university and community members and engages students' families and other community partners as volunteers, learners, and teachers of culturally specific gardening and culinary practices. Parents of students involved in GEPFS are invited to offer cooking demos during the class where they teach students how to cook a recipe relevant to their specific cultural heritage using ingredients harvested from university and community gardens. Students also participate in hands-on gardening and soil science learning activities that demonstrate the use and benefits of organic amendments and water-efficient irrigation practices. GEPFS also involves demonstrations and field trips to showcase how sustainable agricultural practices are being implemented in their region and provide students with practical knowledge and skills. For example, students have engaged in on-campus demonstrations of drought-resistant plants and pollinator-friendly species that integrate biodiversity-enhancing techniques. Students also practice observational nature journaling as a way to creatively reflect on the biodiversity they encounter in both on and off-campus gardens and farms. Educational programs that offer variations on these examples can prepare high school and college students to become ambassadors for food sovereignty, fostering a new generation of environmentally conscious students and advocates for environmental justice.

The ongoing design of GEPFS remains to be determined as alignment with a food sovereignty paradigm requires community control of the food system which includes collaborative, co-construction of agroecological education. While these preliminary findings provide critical insights into the specific design of a forthcoming educational program, what is perhaps most important about these findings is that they indicate opportunities for deeper collaboration between university researchers and community members in the pursuit of food sovereignty. Our aim is not to propose a strict model for agroecological education, but rather to demonstrate how rooting any approach to agroecological education and research in community visions and dreams can uncover immense possibilities for community-based approaches to food sovereignty-oriented inquiry and lead to the co-creation of a more just food system for all.

References

Alaboudi, Khalid A., Ahmed Berhan, and Brodie Graham. "Phytoremediation of Pb and Cd contaminated soils by using sunflower (Helianthus annuus) plant." *Annals of Agricultural Sciences* 63.1 (2018): 123–127.

Awa, Soo Hui, and Hadibarata Tony. "Removal of heavy metals in contaminated soil by phytoremediation mechanism: A review." *Water, Air, & Soil Pollution* 231.2 (2020): 47.

Bentley, M. "An ecological public health approach to understanding the relationships between sustainable urban environments, public health and social equity." *Health Promotion International* (2014):29(3), 528–537.

Borck, C. "I belong here.": Culturally sustaining pedagogical praxes from an alternative high school in Brooklyn. *The Urban Review*. (2020): 52(2), 376–39111.

Borrelli, Pasquale, et al. "Land use and climate change impacts on global soil erosion by water (2015-2070)." *Proceedings of the National Academy of Sciences* 117.36 (2020): 21994–22001.

Brones, A. Karen Washington: It's not a food desert, it's food apartheid. *Guernica* (2018, May 7) https://www.guernicamag.com/karen-washington-its-not-a-food-desert-its-food-apartheid

Cachelin, A., Ivkovich, L., Jensen, P., & Neild, M. Leveraging foodways for health and justice. *Local Environment* (2019): *24*.(5): 417–427. https://doi.org/10.1080/13549839.2019.1585771

Cachelin, A., Joyner, L., Kuttner, P., Rejon Magana, G., Montoya, E., Martinez, J., Taddie, K., Yagüe B., and Banerjee, D. (in press). Embracing the journey: Growing educational pathways for food sovereignty in Antunes, A. and Howard, J. (Eds). *Urban research based action networks*. Meyers Press.

Chancellor's Office Study on Basic Needs. Basic Needs Initiative. (2018). https://basicneeds.sfsu.edu/sites/default/files/documents/CSU%20Basic%20Needs%20Report%202018-%20SF%20State%20Specific%20Data%20Points%20%20%281%29.pdf

Chien, Chihching, et al. "Microbial diversity of soil bacteria in agricultural field contaminated with heavy metals." *Journal of Environmental Sciences* 20.3 (2008): 359–363.

Delta Independence Science Board. "*Water quality science in the Sacramento – San Joaquin Delta, chemical contaminants and nutrients*". Delta Stewardship Council. (2018).

EPA. (2023). https://www.epa.gov/heatislands

Fisher, A. "*Big hunger: The unholy alliance between corporate America and anti-hunger groups*." The MIT Press (2017). https://doi.org/10.7551/mitpress/10987.001.0001

Gao, Si, Hoffman-Krull Kai, and DeLuca Thomas H. "Soil biochemical properties and crop productivity following application of locally produced biochar at organic farms on Waldron Island, WA." *Biogeochemistry* 136.1 (2017): 31–46.

Goldsworthy, Andrew. "Effects of electrical and electromagnetic fields on plants and related topics." in A. G. Volkov (Ed). *Plant electrophysiology: Theory and methods*. Springer Berlin Heidelberg (2006): 247–267.

Gregory, M. M., Leslie, T. W., and Drinkwater, L. E. "Agroecological and social characteristics of New York city community gardens: Contributions to urban food security, ecosystem services, and environmental education." *Urban Ecosystem*. (2015): 19, 763–794.

Guitart, Daniela, Pickering Catherine, and Byrne Jason. "Past results and future directions in urban community gardens research." *Urban Forestry & Urban Greening* (2012): 11(4), 364–373.

Hanak, Ellen, et al. "Water and the future of the San Joaquin Valley." *Public Policy Institute of California* 100 (2019): 3–14.

Havugimana, Erneste, et al. "Soil pollution–major sources and types of soil pollutants." *Environmental Science and Engineering* 11 (2017): 53–86.

Holt-Giménez, E., & Shattuck, A. "Food crises, food regimes and food movements: Rumblings of reform or tides of transformation?" *The Journal of Peasant Studies*, (2011): 38(1), 109–144.

International Panel of Experts on Sustainable Food Systems IPESFS. (2020). *COVID-19 and the crisis in food systems: Symptoms, causes, and potential solutions*. Retrieved from https://ipes-food.org/report/covid-19-and-the-crisis-in-food-systems/

Joyner, L., Yagüe, B., Cachelin, A. "Promoting food systems paradigm shifts through critical reflexivity: Exploring interviews as intervention." *Social Sciences* (2023): 12(5), 280. MDPI AG. Retrieved from http://doi.org/10.3390/socsci12050280

Joyner, L., Yagüe, B., Cachelin, A. & Rose, J. "Farms and gardens everywhere but not a bite to eat? A critical geographic approach to food apartheid in Salt Lake City". *Journal of Agriculture, Food Systems, and Community Development* (2022): 11(2). https://foodsystemsjournal.org

Katz, L. "A Racist Past, a Flooded Future: Formerly Redlined Areas Have $107 Billion Worth of Homes Facing High Flood Risk—25% More Than Non-Redlined Areas." RedFin News (2021) Retrieved from https://www.redfin.com/news/redlining-flood-risk/

Khan, Naser, et al. "Complementing compost with biochar for agriculture, soil remediation and climate mitigation." *Advances in Agronomy* 179 (2023): 1–90.

Kuttner, P. J.(2023) "The right to belong in school: A critical, transdisciplinary conceptualization of school belonging." *AERA Open*, *9*, 23328584231183407.

Laurin, Jeremy T., and Luengen Allison C.. "Assessment of the contribution of utility vault water to surface water pollution." *Environmental Monitoring and Assessment* 191.7 (2019): 467.

Lillis, R. "*Sacramento refuses to clean lead-tainted lawns near gun range, says airport is to blame.*" Center for Environmental Health (2017). Retrieved from https://ceh.org/latest/news-coverage/sacramento-refuses-clean-lead-tainted-lawns-near-gun-range-says-airport-blame/#:~:text=In%20that%20Oct.,of%20airports%20in%20the%20nation

Lin, Brenda B., Philpott Stacy M., and Jha Shalene. "The future of urban agriculture and biodiversity-ecosystem services: Challenges and next steps." *Basic and Applied Ecology* 16.3 (2015): 189–201.

Livesley, Stephen J., et al. "Soil-atmosphere exchange of carbon dioxide, methane and nitrous oxide in urban garden systems: Impact of irrigation, fertiliser and mulch." *Urban Ecosystems* (2010): *13*, 273–293.

Manisalidis, Ioannis, et al. "Environmental and health impacts of air pollution: A review." *Frontiers in Public Health* 8 (2020): 14.

Mann, A. "Education for food sovereignty as transformative ethical practice." *Policy Futures in Education*. (2019): 17(7), 862–877. https://doi.org/10.1177/1478210318816251

Martínez-Torres, M., & Rosset, P. "La Vía Campesina: The birth and evolution of a transnational social movement." *The Journal of Peasant Studies*. (2010): 37(1), 149–175. https://doi.org/10.1080/03066150903498804

Meek, D. The cultural politics of the agroecological transition. *Agriculture and Human Values* 33, 275–290 (2016). https://doi.org/10.1007/s10460-015-9605-z

Nyéléni, N. "Forum for food sovereignty [Declaration]." Sélingué, Mali. (2007). https://nyeleni.org/DOWNLOADS/Nyelni_EN.pdf

Oliver, Margaret A., and Gregory P. J.. "Soil, food security and human health: A review." *European Journal of Soil Science* 66.2 (2015): 257–276.

Osterman, K. F. "Students' need for belonging in the school community." *Review of Educational Research*. (2000): *70*(3), 323–367.

Palik J., and Østby G. "Interventions to improve refugee children's access to education and quality learning: A scoping review of existing impact evaluations." *International Review of Education*. (2023): *69*, 227–247.

Pathak, Tapan B., et al. "Climate change trends and impacts on California agriculture: A detailed review." *Agronomy* 8.3 (2018): 25.

Peña, D.G., Calvo, L., McFarland, P., & Valle, G.R. (2017). *Mexican-Origin Foods, Foodways, and Social Movements: Decolonial Perspectives*. The University of Arkansas Press.

Reese, A. M. "*Black food geographies: Race, self-reliance, and food access in Washington, D.C.*" The University Of North Carolina Press. (2019). https://doi.org/10.5149/northcarolina/9781469651507.001.0001

Sbicca, J. "Growing food justice by planting an anti-oppression foundation: Opportunities and obstacles for a budding social movement." *Agriculture and Human Values*. (2012): *29*(4), 455–466. https://doi.org/10.1007/s10460-012-9363-0

Shaheb, Md Rayhan, Venkatesh Ramarao, and Shearer Scott A. "A review on the effect of soil compaction and its management for sustainable crop production." *Journal of Biosystems Engineering* (2021):*46*, 417–439.

Shepherd, S., et al. "Extremely low frequency electromagnetic fields impair the cognitive and motor abilities of honey bees." *Scientific Reports* 8.1 (2018): 1–9.

Tehrani S. O., Wu S. J., Roberts J. D. "The color of health: Residential segregation, light rail transit developments, and gentrification in the United States." *International Journal of Environmental Research and Public Health* (2019): 16(19), 3683.

Trauger, A. "We *want land to live: Making political space for food sovereignty*." University of Georgia Press. (2017).
Valley Vision. https://www.valleyvision.org/wp-content/uploads/The-Food-System-Resilience-Poll-October-2021_web.pdf Food System Resilience Poll (2021). https://www.valleyvision.org/wp-content/uploads/2021RegionalActionPlan.pdf
Yang, Zhizhou, et al. "Spatiotemporal variation characteristics and source identification of water pollution: Insights from urban water system." *Ecological Indicators* 139 (2022): 108892.

6 Green Space Provision and Equity in the City

Magdalena Biernacka

Introduction

In the face of numerous crises, including pandemics and related diseases (e.g., depression, anxiety disorders), universal access to green spaces such as urban parks and greenways should be available to all residents of urban areas, regardless of their socioeconomic status, ethnic group, age, or gender. Access to green spaces supports people's psychophysical well-being, improving health and quality of life (Enssle and Kabisch 2020; Lee and Maheswaran 2011). This further translates into life satisfaction, greater productivity, and a longer life (Reyes-Riveros et al. 2021). Numerous studies show that physical activity in green spaces reduces stress, improves the cardiovascular system, and thus reduces the incidence of "diseases of civilization" such as obesity, diabetes, mental illnesses, and heart disease (Schipperijn et al. 2017; Hedblom et al. 2019; Callaghan et al. 2021).

Another contemporary threat we face is climate change, which affects not only the well-being of city residents but also the condition and existence of green spaces in urban areas (Onder and Dursun 2010). In regions with a temperate climate, we observe a decrease in snow cover duration and a rise in temperatures during winter and summer. These changes can undoubtedly have a detrimental impact on green spaces; extensive forest areas may be destroyed by droughts or tornadoes, for example, or green spaces could be submerged due to flooding. Therefore, it is crucial to build cities' resilience to climate change and to design green spaces to reduce the negative impacts of climate change (Meerow and Newell 2017). The key is to enhance natural ecosystems and link green spaces through ecological corridors such as linear parks. This supports high biodiversity and offers residents a place to rest. On hot days, these areas become sought-after havens for relaxation and recreation under the shade of trees and other vegetation.

It is crucial to ensure that connections are made between different types of green space; not only between formal and managed green spaces such as parks, squares, or forests but also between so-called informal green spaces (e.g., vacant lands, fields, wastelands). Only combining different green spaces can provide a coherent blue-green infrastructure system that translates into

DOI: 10.4324/9781032631110-7

numerous benefits in the form of ecosystem services (Brears 2018; Zalewski et al. 2012). It is essential to recognize informal green spaces as a valuable resource that provides additional space for recreation as well as enhances biodiversity in the city (Sikorski et al. 2021).

This chapter addresses issues related to the provision of green spaces – why it is not only the existence (availability) of green spaces that matters but also the opportunities to use green spaces (accessibility) and individual perceptions (attractiveness), limitations, and socioeconomic status that shape green space provision and equity. In the next section, we will present the classification of green spaces and blue-green infrastructure, making the distinction between formal and informal green spaces and emphasizing the importance of the latter category. The quantification of green space provision, with a particular focus on the latest approach known as the "3-30-300 rule", is also discussed. Finally, issues related to ecosystem services, with an indication of what actions should be taken so that residents can benefit from as many ecosystem services as possible, are considered. Local case studies have been added to each section to showcase different approaches and highlight current challenges and opportunities.

Green Space Provision: Availability, Accessibility, and Attractiveness

The supply of green space in the city consists of three levels: availability (i.e., the existence of green space), accessibility, and attractiveness. It is crucial to distinguish between these three levels and see that each successive level is based on availability (Biernacka and Kronenberg 2018) (Figure 6.1).

The most basic level of green space provision, availability, refers to the existence of green spaces and is related to various planning decisions that can translate into the appearance of a new green space, the protection of an existing one, or the removal of a green space (e.g., due to new development of roads, buildings, etc.). This is the most basic level that affects whether residents can benefit from a green space. The second level, accessibility, addresses physical accessibility (no fences, time restrictions to enter, fast roads to avoid as there are no pedestrian lanes) and psychological accessibility (i.e., the fact that we are not afraid to use a given green space; we feel safe and welcome there). The last level,

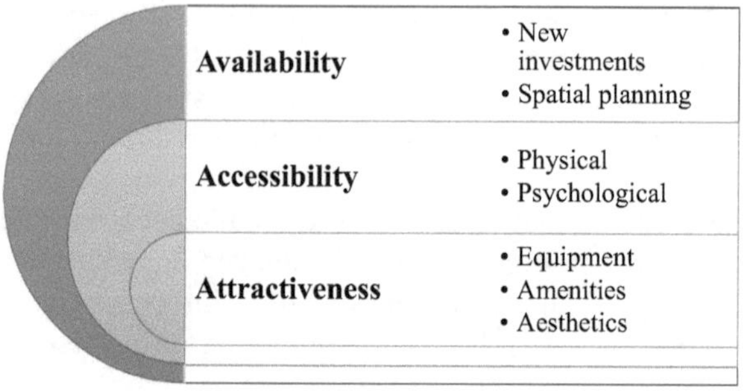

Figure 6.1 Three levels of green space provision. Credit: Biernacka and Kronenberg (2018).

that of attractiveness, often determines whether we want to use the green space because there is adequate equipment (e.g., paths, playgrounds, benches, sports fields) and positive aesthetic qualities (not polluted, degraded, or littered). If the facilities and surroundings meet our preferences, we will use the green space and stay there for extended periods (Figure 6.1). The model of three levels of supply of green spaces places particular emphasis on various barriers that prevent using green spaces, especially those that are institutionally conditioned, based on formal and informal rules in society (Vatn 2007). Similar assumptions are made by Wolff et al. (2022), who also reference institutions but also emphasize perception and psychological aspects. Physical barriers can be encountered both inside and outside of green spaces, such as a lack of equipment, a lack of public transportation, or a lack of parking lots. Institutional barriers include planning decisions and cultural norms. Barriers related to the limitations of a specific person include individual limitations resulting from disability or interpersonal relationships. This "barriers perspective" enables decision-makers to identify specific problems and errors that they should address to increase the supply of green spaces for all groups of residents, especially the most vulnerable (Biernacka, Łaszkiewicz, and Kronenberg 2022).

Another approach related to the flow of benefits from green spaces to residents is a concept based on three pillars: perception, institutions, and infrastructure (Andersson et al. 2019). It combines elements related to the barrier approach but also adds elements related to how green spaces are perceived by residents. The key step here is to add the perceptual element, as residents differ in their perception and appreciation of green spaces. What is desirable and preferred for one person (e.g., crowds and loud entertainment events) may be a repulsive factor for another.

When analyzing the three levels of green space provision and various barriers, equality is fundamental, that is, who has access to green spaces and to what extent this access may be limited or difficult. The first factor that may influence the use of green spaces is wealth. Research shows that more affluent households with a higher socioeconomic status live in urban neighborhoods and districts with more greenery, while poorer residents tend to have more limited access (Rigolon 2016; Rigolon et al. 2018). It is also worth mentioning that some parks and special gardens are fenced, and entry fees (sometimes prohibitively high) are charged (Biernacka and Kronenberg 2018). Such high entrance fees may exclude especially large families and the elderly (with low incomes) from using a given green space (e.g., zoological gardens, botanical gardens). Transportation may also be an issue, as transit options may be limited and some green spaces can only be reached by car (Mohamed, Kronenberg, and Łaszkiewicz 2022). Less-wealthy households who cannot afford to maintain a car, or older adults who are unable to drive for health reasons, may also be excluded here. One of the more vulnerable groups who may have difficulty accessing green spaces is that with disabilities. Green spaces are often unsuitable for their needs and lack essential equipment such as adapted toilets, paved surfaces, and appropriate information for the visually impaired (Corazon et al. 2019; Dahlberg et al. 2022).

An extreme case of lack of equality in access to services, infrastructure, and green spaces can be observed in Cape Town, South Africa (https://unequalscenes.com/cape-town) (2024). Observing the photos from a bird's-eye view, we can see the boundaries between rich villas and green estates where wealthier residents live and less green poorer districts (including slums). The poorer communities are characterized by an extreme lack of greenery and are often found in unfavorable or undesirable locations (e.g., directly bordering a cemetery, flood-prone areas). Such a view is not uncommon, and the lack of equal access to green spaces can be observed in many countries (e.g., Brazil, Mexico, Pakistan); this translates into limited opportunities to benefit from ecosystem services which directly affects the health and quality of life of disadvantaged residents (Rigolon 2016; Rigolon et al. 2021).

Classification of Urban Green Spaces and Green Infrastructure Types

Urban green spaces can be classified in various ways (e.g., in terms of size, functions performed, and property rights), with a particularly important division between formal and informal green spaces. Formal green spaces are publicly accessible, designed and managed by local authorities, and informal green spaces are all green spaces that are not formally counted as green space, but which contain trees, shrubs, grasses, and generally biologically active areas. Such formal green spaces may include, for example, parks, forests, cemeteries, and allotment gardens, while informal green spaces may include wastelands, brownfields, meadows, private gardens, or greenery around various institutions. Moreover, we can introduce green infrastructure in places where it is not possible to create a larger park or green square in the form of green facades, roofs, or rain gardens. A broader characterization of the different types of green spaces and examples of green infrastructure is presented in Table 6.1 (Biernacka et al. 2023; Nochian et al. 2015; Rupprecht et al. 2015; Kim, Rupprecht, and Furuya 2018).

Formal green spaces are usually extensively maintained; grass is cut, leaves are raked, and shrubs are pruned. A particular case of a formal green space that has retained its wild character is the Drwinka River Park in Krakow, Poland. The Drwinka River Park is an example of a place that should be more formalized to preserve its wild character and prevent the introduction of large-scale developments. It is also an example of how grassroots initiatives can contribute to preserving the city's green spaces and how the voice of residents can make a difference. In this case, greater formalization of the park, covering it with local zoning plans and appropriate transformations, can ensure its protection, prevent it from being degraded, and introduce significant anthropological transformations. This has been advocated by the Drwinka Association, which has several dozen members who work vigorously to maintain the park's health by organizing various educational projects involving walks and educational materials, creating publications, cleaning the park, and staging protests when necessary. Another insightful example of a bottom-up initiative and cooperation between city actors is Biomass Park Hugo, developed on the site of a former mine in Gelsenkirchen, Germany (for details, see Case Study 1).

Table 6.1 Types of green space and green infrastructure in urban areas.

Formal green spaces and green infrastructure – examples		*Characteristic*
Parks, green squares	• Community parks • Linear parks • Local parks • Mini-parks • Pocket parks • Public parks • Urban parks • Waterfront parks	Parks are usually public and open green spaces managed by the city. They can serve a variety of functions depending on their size and shape; e.g., linear parks provide a natural ecological corridor for people and animals, while pocket parks complement the greenery in city centers. They are most often well maintained and equipped with park infrastructure such as benches, litter bins, and recreational amenities such as playing fields and playgrounds. Most typical place for recreation and physical activity by city residents.
Neighborhood green spaces	• Fenced housing estates • Housing communities	Neighborhood green spaces are most often well-kept and located around apartment blocks or suburban subdivisions. Small infrastructure exists, such as benches, playgrounds, and lighting. It can be accessible to all residents or be fenced. It is the closest place of recreation for many city residents, especially the elderly and families with children.
Special gardens	• Botanical garden • Palm house • Private parks owned by a private owner, company • Zoological garden	Special gardens, such as the botanical garden, are city-managed (sometimes privately managed) places with collections of exotic plants or animals. They are highly maintained green spaces with small park infrastructure and other amenities (toilets, cafés). Usually, there is a fee to enter them. An interesting example is Aflopark, the largest private garden in central Poland covering an area of 6 ha, which has hundreds of varieties of roses and lilies in its collection.
Forests	• Public forests • Private forests	Large green spaces with a high proportion of dense trees, usually have a wild character. They may be equipped with paths, benches, and sometimes other recreational amenities. Private forests are most often unfenced and used for economic purposes, and access to their area is prohibited to people who do not own the land.
Cemeteries	• Municipal • Religious denominations	A place for burial, old cemeteries are home to numerous trees, shrubs, and also wild animals.

(*Continued*)

Table 6.1. (Continued)

Formal green spaces and green infrastructure – examples		*Characteristic*
Allotment gardens, community gardens	• Allotment gardens in perpetuity • Community gardens • Social gardens	Allotment gardens are specially designed and subdivided into smaller plots, a place of leisure and recreation for the city's residents, where they can grow vegetables and fruit for their own use. Usually well-maintained green spaces that are accessible to a small group of people. In turn, community gardens are created and maintained by the local community and are used for growing vegetables and herbs, but also for community integration and social events.
Street greenery	• Flower meadows • Infiltration trenches • Ornamental planting along traffic routes • Street trees and shrubs	The greenery along the streets is usually managed by the city and provides a protective element, e.g., against noise and pollution, but also has a decorative function and can absorb part of the rainwater, thus providing regulating ecosystem services.
Vertical greenery	• Green viaducts • Green walls, facades • Vertical gardens	Vertical greenery can be introduced in areas where there is a lack of planting or space, especially in city centers.
Horizontal greenery	• Green bus stops • Green roofs	Green roofs and green bus stops should be introduced where there is a lack of green space; in addition to their aesthetic function, they can reduce excessive surface runoff and improve the quality of the urban environment.
Small-scale greenery	• Parklets • Pot planting • Rain gardens	Introducing miniature green spaces is very important as it positively influences the perception of space and improves the quality of the environment.
Informal green spaces – examples		*Characteristic*
Brownfields, vacant lots	• Underutilized industrial, commercial, and housing	They are areas that are mostly undeveloped and abandoned and provide a place for spontaneous plant vegetation and animal habitation. They can constitute the so-called fourth nature in a city. They can be dangerous for visitors because of the remains of old buildings and architectural structures.

Cultivated lands: current and former	• Fields • Meadows • Orchards • Fallow lands	Agricultural land in the city often becomes wasteland, e.g., old orchards. Some areas are converted into new housing developments. Former agricultural lands or fallow lands are not managed in any way and are sites of spontaneous vegetation.
Waterside green spaces	• Greenery located in the river valley, by the lake or canal	The greenery accompanying a river, stream, canal, or lake, which is often quite wild in nature, can provide a habitat for animals as well as a recreational space for residents.
Green spaces associated with the activities of various institutions	• Religious denominations • Industry • Offices	The green spaces around the institution are usually well maintained and landscaped (especially around institutions that care about image and appearance) and can serve a variety of functions, although most often they are used for recreation by the institution's staff or visitors (e.g., around office buildings or banks). The greenery around industry can additionally act as a buffer and protection against noise, however, it sometimes becomes a source of waste, a wild landfill, e.g., at car repair stations.
Green spaces accompanying educational facilities	• School greenery • Kindergarten greenery • Universities • Didactic gardens	Greenery is an element of the surroundings of a given institution; it serves a recreational function as a place for physical activity (sport fields) but also for play (playgrounds) and education (e.g., educational or experimental gardens located at universities).
Protective green spaces	• Green spaces around railway tracks • Green spaces around nuisance facilities, e.g., airports, sewage treatment plants, landfill sites	Constitutes a protective zone between residents and nuisance facilities. Thanks to existing trees, noise, air pollution, and unpleasant odors can be reduced. Additionally, it is a form of aesthetic protection.
Private gardens	• Backyards • Front yards	These are green spaces that have a private owner. The gardens around houses are usually well kept and are used for recreation by the household members. Sometimes they have ponds, decorative plantings, or small infrastructure for children to play.

Case Study 1 – Biomass Park Hugo

To recultivate the area and restore its ecological potential, a biomass plantation – the Biomass Park Hugo – was created in the relic mine "Hugo 2/5/8" in the City of Gelsenkirchen, Germany, accompanied by a community garden. When designing the plantation, the possibility of its multifunctional use was achieved through ecological education, scientific research, nature protection, recreation, and the creation of interpersonal bonds. This facility is an economically and ecologically interesting element of the landscape of Gelsenkirchen, as well as a kind of compensation for the intensive industrial use of this area over the last 130 years. The central part of the site serves as a "landscape laboratory" with a test area for energy plants (grown for biomass), such as dogwood or paulownia, and an educational area used by nearby schools and kindergartens. In the community garden, there are fruit, vegetable, and herb crops, insect houses, picnic areas, and places for small water retention. The area is ideal for the reproduction of the natterjack toad and serves as a habitat for many different plant species and insects. It is a place willingly visited by local residents and used for recreation as well as growing crops for local consumption.

Informal green spaces are often overlooked, especially in official statistics, which mainly refer to formal green cover. However, a growing number of studies show that informal green spaces make up a significant percentage of green space in many cities (Feltynowski et al. 2018). The categorization of these spaces in Lodz, Poland, for example, clearly shows that informal green spaces dominate, so taking them into account in formal city plans and projects is key to positively influencing the quality of life in the city (for details, see Case Study 2).

Case Study 2 – Green spaces in Jasień and Karolewka river valleys

The Jasień and Karolewka river valleys are located in Lodz, Poland, a post-industrial city that experienced intensive development in the 19th century. This development led to environmental pollution and the transformation of many watercourses into underground channels. One of these canal rivers is the Jasień River, which runs through the entire city, starting in the outskirts, running through the city center, and ending again on the outskirts of town. A number of green spaces can be identified around these river channels, which have been classified in detail (Figure 6.2). Researchers have identified 17 categories of green spaces with formal green spaces representing only 24% (six categories). Along the urban-rural gradient within these river valleys, informal green spaces dominate (76%; 11 categories). Moreover, more than half of the informal green spaces are privately owned and dominate in low-density suburbs (Biernacka et al. 2023). What is important is that such areas are often overlooked by decision-makers and not included in various strategies and planning documents. They are undervalued, and their high recreational potential is not perceived by many; in fact, they are often considered a problem. In such a case, cooperation between private owners and the city is needed to make these green spaces available to residents. This may involve putting minimal infrastructure (e.g., paths, benches) and amenities in place so that these spaces can serve as a great alternative to formal and designed green spaces like urban parks.

Urban decision-makers should begin to recognize informal green spaces as a valuable resource that can provide many benefits to residents and influence a city's resilience to climate change.

Among the most important of these benefits include:

- Additional green space for residents as an alternative to formal green spaces, which may be insufficient in number or may be crowded (Sikorska et al. 2020; Pietrzyk-Kaszyńska, Czepkiewicz, and Kronenberg 2017).
- Informal green spaces provide habitats for various plant and animal species and can be more biodiverse than parks or managed green squares. It constitutes the so-called fourth nature in the city (Sikorski et al. 2021).
- It can provide a natural link (ecological corridor) between different green spaces; for example, a formal park and an informal waterside area, thus fostering a network of blue-green infrastructure, additionally providing many ecosystem services (Włodarczyk-Marciniak, Sikorska, and Krauze 2020).

Quantification of Green Space Provision

Various indicators have been introduced to assess to what extent residents can benefit from different green spaces (usually formal) (Jia et al. 2021). One such measure is to designate a buffer around a residence and calculate the quantity or density of open green spaces within a given distance, assuming the resident can use it (Su et al. 2019). Alternatively, the amount of green space can be calculated per capita (i.e., per person); this way, we can determine the minimum number of square meters of green space per inhabitant (WHO 2010). We can also designate service areas and determine how far or how much time it takes for residents to reach a given green space (Biernacka, Łaszkiewicz, and Kronenberg 2022). There is also renewed interest in neighborhood walkability and walking distance (Schipperijn et al. 2010); that is, ideally we should spend less than 15 minutes on foot to reach a green space from home, which is in line with the concept of the 15-minute city and compact city paradigm.

Figure 6.2 Informal green spaces around the Jasień River in Lodz, Poland. Credit: Jakub Kronenberg.

The latest rule for assessing green space provision combines the view of greenery from home windows, the area of greenery in the immediate vicinity of home, and the distance from home to the nearest green space; it is called the 3-30-300 rule (Konijnendijk 2021). According to the rule, each resident should have a view of at least three trees from the window (especially for mental health), and the neighborhood should be covered by 30% green space (with positive effects on microclimate, air purification, and temperature). Also, the distance from the house to the most immediate green space used for recreation should be no more than 300 meters. This principle combines different measures and shows straightforwardly how important it is to be able to both view and physically utilize green spaces to benefit from as many ecosystem services as possible.

A collaboration between an international project team and the city of Turku, Finland resulted in the development of the blue-green factor (https://www.turku.fi/en/blue-green-factor; City of Turku, n.d.). This is a tool used by designers and planners, which, by adding up the components of greenery, allows the ecological efficiency of a plot to be calculated. The initial data to be entered includes information about the plot, planned buildings, and target values. The amount of vegetation, paving, and rainwater structure, on the other hand, can be found in the special calculations table (excel forms). The resulting data are the green factor, as well as the proportion of impervious surface and the volume of rainwater retention structures. The creation of green courtyards is key in terms of improving the quality of life of residents, but they also have an impact on balancing the local climate (i.e., regulation of humidity and windiness) and affect the quality of discharge from water bodies. The blue-green factor is used successfully in many Finnish cities.

The most basic measure of green space in Poland is the biologically active surface, which refers to the surface area where natural vegetation can occur. In many cities, this measure has been extended to include various biologically active elements such as trees, green walls and roofs, and planting in pots or balconies. One such measure is the Green Space Index proposed in Poland (Lodz), which considers various elements of green and blue infrastructure (see Case Study 3). These measures assess the extent to which a plot or area is ecologically active and can provide specific ecosystem services (Elderbrock et al. 2020).

Case Study 3 – Green Space Index

In 2018, the Municipal Planning Office in Lodz, Poland proposed a Green Space Index as an extension of the biologically active surface with additional elements such as trees, shrubs, green walls, or ponds. Three groups of actively green elements appear in the indicator, each assigned a corresponding coefficient value. By identifying the individual elements and counting the indicators assigned to them, we can determine the actual biological activity of the plot. This is an exciting proposal that should be included in planning records. Similar proposals have been put forward in other cities, such as the Green Space Factor in Malmö and the Seattle Green Factor. They operate similarly to the Polish indicator but consider additional elements such as planting flowers on balconies or the special adaptation of green roofs for growing plants or vegetables. The use of such indicators and their legal sanctioning is crucial to improving the city's microclimate, increasing water retention, and developing incentives for residents to invest in biologically active elements on their plots of land, such as through financial subsidies and tax relief.

Ecosystem Services – Why Do Urban Residents Need Something More Than a Plain Lawn?

City dwellers should have access to various environmental benefits of green spaces known as ecosystem services, which can be divided into four main groups: provisioning, supporting, regulatory, and cultural services (Haines-Young and Potschin 2010). Depending on the type of green space, we can talk about different types of ecosystem services. For example, informal green spaces provide essential regulatory services, a habitat for many plant and animal species, and contribute to the microclimate and temperature of the city (Sikorski et al. 2021). Of course, formal green spaces such as parks or landscaped gardens (e.g., botanical gardens) also provide services (e.g., thermal comfort, noise reduction), but in their case the cultural function dominates; they are primarily designed to provide a place for recreation, leisure, human relations, and spiritual and aesthetic experiences (Buchel and Frantzeskaki 2015).

It is therefore crucial to maintain different types of green spaces that provide different types of ecosystem services and support biodiversity. Furthermore, such formal and informal green spaces should be interconnected so that additional benefits, such as the safe movement of animals, can be maintained. Such ecological corridors or networks also benefit people, as they can walk from one green space to another without being exposed to noise, air pollution, or other negative externalities common in the city (Letter and Jäger 2020; Margaritis and Kang 2017).

Additionally, green corridors can serve to protect natural ecosystems, promote the integrity of the water cycle, and be designed in a sustainable way that promotes the city's resilience to climate change. Examples of such green corridors include the Cerros Orientales Ecological and Recreational Corridor in Bogotá, Colombia or the Rutas Naturbanas project in San José, Costa Rica (Ghisleni 2023). Green corridors should serve residents, but also have a positive impact on the environment and be integrated so as not to disturb the natural landscape.

Summary

The provision of green spaces includes availability, accessibility, and attractiveness, and each is important; it is worth remembering that the existence of green spaces is paramount, but whether these spaces are available and attractive to residents also has a significant impact. Barriers that affect the supply and accessibility of green spaces may result from institutional conditions, as well as personal perceptions and individual limitations (e.g., disability, old age). Therefore, a complex analysis of residents' needs can be incredibly useful. Such analyses should consist of both quantitative (e.g., area of parks, amount of equipment, transport connections) and qualitative measures, the latter of which will enable a determination of the actual preferences and needs of users. It is also crucial to recognize the distribution of the most vulnerable groups (those with low socioeconomic status, children, older adults, and the disabled) and design new green spaces or modify existing ones to make them as accessible and attractive as possible.

We can understand the categories of green spaces in the city in different ways. However, it is worth noting that we can distinguish several types of green spaces (formal, e.g. parks, allotment gardens, special gardens, or informal, e.g., wasteland, fields), as well as green infrastructure, which is often integrated into buildings, etc. Each of these elements is essential and provides a range of ecosystem services, particularly informal green spaces, whose role in ensuring biodiversity and contributions to the resilience of cities is often underestimated.

Many indicators and measures have been developed to suggest how much greenery we need and what number/coverage of trees and other vegetation will be optimal for a high quality of life. These indices and measures are particularly useful for planners as they help determine optimal values related to the presence of blue-green infrastructure, which translates into well-functioning urban ecosystems to the benefit of all residents. Additionally, the need to ensure ecological connectivity by creating green corridors between different green spaces is essential for all living organisms, humans as well as plants and animals. By properly assessing, monitoring, and continuously improving the equitable provision of both formal and informal green space within the city, we can ensure a healthier environment for ourselves and generations to come.

References

Andersson, Erik, Johannes Langemeyer, Sara Borgström, Timon McPhearson, Dagmar Haase, Jakub Kronenberg, David N. Barton, et al. 2019. "Enabling Green and Blue Infrastructure to Improve Contributions to Human Well-Being and Equity in Urban Systems." *BioScience* 69 (7): 566–74. https://doi.org/10.1093/biosci/biz058

Biernacka, Magdalena, and Jakub Kronenberg. 2018. "Classification of Institutional Barriers Affecting the Availability, Accessibility and Attractiveness of Urban Green Spaces." *Urban Forestry & Urban Greening* 36 (December): 22–33. https://doi.org/10.1016/j.ufug.2018.09.007

Biernacka, Magdalena, Edyta Łaszkiewicz, and Jakub Kronenberg. 2022. "Park Availability, Accessibility, and Attractiveness in Relation to the Least and Most Vulnerable Inhabitants." *Urban Forestry & Urban Greening* 73 (July): 127585. https://doi.org/10.1016/j.ufug.2022.127585

Biernacka, Magdalena, Jakub Kronenberg, Edyta Łaszkiewicz, Piotr Czembrowski, Vahid Amini Parsa, and Daria Sikorska. 2023. "Beyond Urban Parks: Mapping Informal Green Spaces in an Urban–Peri-Urban Gradient." *Land Use Policy* 131 (August): 106746. https://doi.org/10.1016/j.landusepol.2023.106746

Brears, Robert C. 2018. *Blue and Green Cities: The Role of Blue-Green Infrastructure in Managing Urban Water Resources*. UK: Springer.

Buchel, Sophie, and Niki Frantzeskaki. 2015. "Citizens' Voice: A Case Study about Perceived Ecosystem Services by Urban Park Users in Rotterdam, the Netherlands." *Ecosystem Services* 12 (April): 169–77. https://doi.org/10.1016/j.ecoser.2014.11.014

Callaghan, A., G. McCombe, A. Harrold, C. McMeel, G. Mills, N. Moore-Cherry, and W. Cullen. 2021. "The Impact of Green Spaces on Mental Health in Urban Settings: A Scoping Review." *Journal of Mental Health* 30 (2): 179–93. https://doi.org/10.1080/09638237.2020.1755027

City of Turku. n.d. "Blue-green factor." Accessed May 15, 2024. https://www.turku.fi/en/blue-green-factor

Corazon, Sus Sola, Marie Christofferen Gramkov, Dorthe Varning Poulsen, Victoria Linn Lygum, Gaochao Zhang, and Ulrika Karlsson Stigsdotter. 2019. "I Would Really like to Visit the Forest, but It Is Just Too Difficult: A Qualitative Study on Mobility Disability and Green Spaces." *Scandinavian Journal of Disability Research* 21 (1): 1–13. https://doi.org/10.16993/sjdr.50

Dahlberg, Annika, Sara Borgström, Max Rautenberg, and Nienke Sluimer. 2022. "A Nearby Park or Forest Can Become Mount Everest. Access to Urban Green Areas by People in Wheelchair from an Environmental Justice Perspective: A Stockholm Case." In *Whose Green City?: Contested Urban Green Spaces and Environmental Justice in Northern Europe*, edited by Bianka Plüschke-Altof and Helen Sooväli-Sepping, 19–40. Sustainable Development Goals Series. Cham: Springer International Publishing. https://doi.org/10.1007/978-3-031-04636-0_2

Elderbrock, Evan, Chris Enright, Kathryn A. Lynch, and Alexandra R. Rempel. 2020. "A Guide to Public Green Space Planning for Urban Ecosystem Services." *Land* 9 (10): 391. https://doi.org/10.3390/land9100391

Enssle, Friederike, and Nadja Kabisch. 2020. "Urban Green Spaces for the Social Interaction, Health and Well-Being of Older People— An Integrated View of Urban Ecosystem Services and Socio-Environmental Justice." *Environmental Science & Policy* 109 (July): 36–44. https://doi.org/10.1016/j.envsci.2020.04.008

Feltynowski, Marcin, Jakub Kronenberg, Tomasz Bergier, Nadja Kabisch, Edyta Łaszkiewicz, and Michael W. Strohbach. 2018. "Challenges of Urban Green Space Management in the Face of Using Inadequate Data." *Urban Forestry & Urban Greening* 31 (April): 56–66. https://doi.org/10.1016/j.ufug.2017.12.003

Camilla Ghisleni, "What are Green Corridors?" Arch Daily, January 16, 2023. https://www.archdaily.com/995047/what-are-green-corridors

Haines-Young, Roy, and Marion Potschin. 2010. "*Proposal for a Common International Classification of Ecosystem Goods and Services (CICES) for Integrated Environmental and Economic Accounting*." Copenhagen: European Environment Agency.

Hedblom, Marcus, Bengt Gunnarsson, Behzad Iravani, Igor Knez, Martin Schaefer, Pontus Thorsson, and Johan N. Lundström. 2019. "Reduction of Physiological Stress by Urban Green Space in a Multisensory Virtual Experiment." *Scientific Reports* 9 (1): 10113. https://doi.org/10.1038/s41598-019-46099-7

Jia, Peng, Xinxi Cao, Hongxi Yang, Shaoqing Dai, Pan He, Ganlin Huang, Tong Wu, and Yaogang Wang. 2021. "Green Space Access in the Neighbourhood and Childhood Obesity." *Obesity Reviews* 22 (S1): e13100. https://doi.org/10.1111/obr.13100

Kim, Minseo, Christoph D. D. Rupprecht, and Katsunori Furuya. 2018. "Residents' Perception of Informal Green Space—A Case Study of Ichikawa City, Japan." *Land* 7 (3): 102. https://doi.org/10.3390/land7030102

Konijnendijk, Cecil. (2021) "The 3-30-300 Rule for Urban Forestry and Greener Cities". *Biophilic Cities Journal* 4 (2).

Lee, A. C. K., and R. Maheswaran. 2011. "The Health Benefits of Urban Green Spaces: A Review of the Evidence." *Journal of Public Health* 33 (2): 212–22. https://doi.org/10.1093/pubmed/fdq068

Letter, Chiara, and Georg Jäger. 2020. "Simulating the Potential of Trees to Reduce Particulate Matter Pollution in Urban Areas throughout the Year." *Environment, Development and Sustainability* 22 (5): 4311–21. https://doi.org/10.1007/s10668-019-00385-6

Margaritis, Efstathios, and Jian Kang. 2017. "Relationship between Green Space-Related Morphology and Noise Pollution." *Ecological Indicators* 72 (January): 921–33. https://doi.org/10.1016/j.ecolind.2016.09.032

Meerow, Sara, and Joshua P. Newell. 2017. "Spatial Planning for Multifunctional Green Infrastructure: Growing Resilience in Detroit." *Landscape and Urban Planning* 159 (March): 62–75. https://doi.org/10.1016/j.landurbplan.2016.10.005

Mohamed, Abdelbaseer A., Jakub Kronenberg, and Edyta Łaszkiewicz. 2022. "Transport Infrastructure Modifications and Accessibility to Public Parks in Greater Cairo." *Urban Forestry & Urban Greening* 73 (July): 127599. https://doi.org/10.1016/j.ufug.2022.127599

Nochian, Ashkan, Osman Mohd Tahir, Suhardi Maulan, and Mehdi Rakhshandehroo. 2015. "A Comprehensive Public Open Space Categorization Using Classification System for Sustainable Development of Public Open Spaces." *Alam Cipta (International Journal on Sustainable Tropical Design Research and Practice)* 8 (December): 29–40.

Onder, Serpil, and Sukru Dursun. 2010. "Global Climate Changes and Effects on Urban Climate of Urban Green Spaces." *International Journal of Thermal and Environmental Engineering* 3 (1): 37–41. https://doi.org/10.5383/ijtee.03.01.006

Pietrzyk-Kaszyńska, Agata, Michał Czepkiewicz, and Jakub Kronenberg. 2017. "Eliciting Non-Monetary Values of Formal and Informal Urban Green Spaces Using Public Participation GIS." *Landscape and Urban Planning* 160 (April): 85–95. https://doi.org/10.1016/j.landurbplan.2016.12.012

Reyes-Riveros, Rosa, Adison Altamirano, Francisco De La Barrera, Daniel Rozas-Vásquez, Lorena Vieli, and Paula Meli. 2021. "Linking Public Urban Green Spaces and Human Well-Being: A Systematic Review." *Urban Forestry & Urban Greening* 61 (June): 127105. https://doi.org/10.1016/j.ufug.2021.127105

Rigolon, Alessandro. 2016. "A Complex Landscape of Inequity in Access to Urban Parks: A Literature Review." *Landscape and Urban Planning* 153 (September): 160–69. https://doi.org/10.1016/j.landurbplan.2016.05.017

Rigolon, Alessandro, Matthew H. E. M. Browning, Kangjae Lee, and Seunguk Shin. 2018. "Access to Urban Green Space in Cities of the Global South: A Systematic Literature Review." *Urban Science* 2 (3): 67. https://doi.org/10.3390/urbansci2030067

Rigolon, Alessandro, Matthew H. E. M. Browning, Olivia McAnirlin, and Hyunseo Yoon. 2021. "Green Space and Health Equity: A Systematic Review on the Potential of Green Space to Reduce Health Disparities." *International Journal of Environmental Research and Public Health* 18 (5): 2563. https://doi.org/10.3390/ijerph18052563

Rupprecht, Christoph, Jason A. Byrne, Jenni G. Garden, and Jean-Marc Hero. 2015. "Informal Urban Green Space: A Trilingual Systematic Review of Its Role for Biodiversity and Trends in the Literature." *Urban Forestry & Urban Greening* 14 (4): 883–908. https://doi.org/10.1016/j.ufug.2015.08.009

Schipperijn, Jasper, Ola Ekholm, Ulrika K. Stigsdotter, Mette Toftager, Peter Bentsen, Finn Kamper-Jørgensen, and Thomas B. Randrup. 2010. "Factors Influencing the Use of Green Space: Results from a Danish National Representative Survey." *Landscape and Urban Planning* 95 (3): 130–37. https://doi.org/10.1016/j.landurbplan.2009.12.010

Schipperijn, Jasper, Ester Cerin, Marc A Adams, Rodrigo Reis, Graham Smith, Kelli Cain, Lars B Christiansen, et al. 2017. "Access to Parks and Physical Activity: An Eight Country Comparison." *Urban Forestry & Urban Greening* 27 (October): 253–63. https://doi.org/10.1016/j.ufug.2017.08.010

Sikorska, Daria, Edyta Łaszkiewicz, Kinga Krauze, and Piotr Sikorski. 2020. "The Role of Informal Green Spaces in Reducing Inequalities in Urban Green Space Availability to Children and Seniors." *Environmental Science & Policy* 108: 144–54. https://doi.org/10.1016/j.envsci.2020.03.007

Sikorski, Piotr, Beata Gawryszewska, Daria Sikorska, Jarosław Chormański, Axel Schwerk, Agata Jojczyk, Wojciech Ciężkowski, et al. 2021. "The Value of Doing Nothing – How Informal Green Spaces Can Provide Comparable Ecosystem Services to Cultivated Urban Parks." *Ecosystem Services* 50 (August): 101339. https://doi.org/10.1016/j.ecoser.2021.101339

Su, Jason G., Payam Dadvand, Mark J. Nieuwenhuijsen, Xavier Bartoll, and Michael Jerrett. 2019. "Associations of Green Space Metrics with Health and Behavior Outcomes at Different Buffer Sizes and Remote Sensing Sensor Resolutions." *Environment International* 126 (May): 162–70. https://doi.org/10.1016/j.envint.2019.02.008

Unequal Scenes. 2024. "Cape Town." Accessed May 17, 2024. https://unequalscenes.com/cape-town

Vatn, Arild. 2007. *Institutions and the Environment*. UK: Edward Elgar Publishing.

WHO. 2010. "Urban Planning, Environment and Health: From Evidence to Policy Action." 2010. http://www.euro.who.int/en/health-topics/environment-and-health/Housing-and-health/publications/2010/urban-planning,-environment-and-health-from-evidence-to-policy-action

Włodarczyk-Marciniak, Renata, Daria Sikorska, and Kinga Krauze. 2020. "Residents' Awareness of the Role of Informal Green Spaces in a Post-Industrial City, with a Focus on Regulating Services and Urban Adaptation Potential." *Sustainable Cities and Society* 59 (August): 102236. https://doi.org/10.1016/j.scs.2020.102236

Wolff, Manuel, André Mascarenhas, Annegret Haase, Dagmar Haase, Erik Andersson, Sara Borgström, Jakub Kronenberg, Edyta Łaszkiewicz, and Magdalena Biernacka. 2022. "Conceptualizing Multidimensional Barriers – A Framework for Assessing Constraints in Realizing Recreational Benefits of Urban Green Spaces." *Ecology and Society* 27 (2): 17. https://doi.org/10.5751/ES-13180-270217

Zalewski, Maciej, Wagner Iwona, W. Frątczak, J. Mankiewicz-Boczek, and P. Parniewski. 2012. "Blue-Green City for Compensating Global Climate Change." *The Parliament Magazine* 350: 2–3.

7 Public Space and Environmental Justice in High-Density Cities

Bo-sin Tang and Anvar Mukhamedjanov

Introduction

Public space is crucial in a livable city. It is said to be the "soul" of a city that reveals its quality of life and the character of its people (Tang & Wong, 2021). All great cities have put a lot of attention into the planning and development of public spaces. Despite all these efforts, however, the outcomes are still far from satisfactory. The *World Cities Report 2022* has highlighted at least two major challenges (UN-Habitat, 2022). First, climate change remains one of the greatest global threats to humanity. Urban planning must take urgent action to address climate change. The provision of blue and green infrastructure in the public spaces of cities is seen as instrumental in restoring urban ecosystems. Second, rising inequality in urban society remains unabated. Some disadvantaged populations bear a disproportionate burden of environmental threats. Providing adequate, green, accessible, and inclusive public spaces to all urban dwellers is essential to break this vicious cycle. The recent pandemics have vividly demonstrated that no one is safe unless everyone is safe.

This chapter explores the relationship between public space and environmental equity in the context of a high-density urban built environment. Public space is, undoubtedly, an important component in the field of urban planning. Public space sounds like common sense to all, but it is in fact an elusive concept which could appear in various physical forms and have varying, sometimes contradictory, meanings to different people. In professional practices, urban planners often see public spaces in terms of function, supply, design, and management. The spatial distribution of public spaces is considered mainly in terms of geographical location, physical quantity, accessibility, and facility, rather than environmental justice. Public spaces can play a key role in influencing urban microclimate and ameliorating the uneven distribution of heat stress across neighborhoods in a city. Using the compact high-density city of Hong Kong as an example, this chapter demonstrates how urban planning is both the cause and the solution to these challenges.

DOI: 10.4324/9781032631110-8

Conceptual Ambiguity: Public Spaces or Public Open Spaces

Cities comprise a mosaic of public and private spaces. However, the boundary between the two is neither clear-cut nor straightforward. Public spaces can be a mixture of natural or seminatural features such as forests, rivers, or beaches, and certain man-made environments such as streets, civic squares, or simply the leftover space between buildings. Thus, landscaping is not necessarily a defining feature, and some public spaces can be mostly or fully "grey" with hard cement surfaces or pavements. On the other hand, public spaces may be government owned, managed, and made freely accessible to all, such as urban parks, public sports grounds, or countryside trails. They can also be privately developed, such as open street cafés, private golf courses, or theme parks, and be offered to those "public" who can afford to pay for access. Social inclusivity of public spaces is thus relative rather than absolute. Based on function, perception, and ownership of space types, Carmona (2010) developed a typology of 20 urban spaces in contemporary cities but concluded that a clear categorization is difficult because many spaces are ambiguous in terms of ownership and publicness.

The terms public open spaces (POS) and public spaces are often used synonymously. Though POS are commonly associated with recreation, amenity, and public use, they face a similar issue of conceptual ambiguity due to the diversity of POS in spatial form, design, ownership, management, and public exclusion. For example, depending on one's own classification, POS may include a wide range of public or semipublic parks, outdoor recreation grounds, pedestrianized streets, promenades, landscaped areas, and so on. Like public spaces, POS are not necessarily green, entirely public, nor fully accessible to all. Therefore, any examination of the effects of POS has to be extremely cautious as to ascertain what specific characteristics of the POS under study are causing the impacts (Mukhamedjanov et al., 2021).

Much of the literature has shown that quality POS contributes positively to cities as part of the blue and green infrastructure (e.g., Li et al., 2022; Farkas et al., 2023). First, POS can encourage more physical activities among the population, leading to better physical health and a lower risk of chronic diseases such as obesity, diabetes, and heart disease (Lai et al., 2021; Wan et al., 2021). Green spaces such as parks, gardens, and playgrounds provide opportunities for exercise, walking, cycling, and other recreational activities. Second, access to nature and greenery in POS can enhance mental health by reducing stress, anxiety, and depression and improving overall well-being (Dzhambov et al., 2021). Some studies have shown that spending time in green environments can improve mood, foster social cohesion, and build a sense of place through community activities and events (Browning et al., 2022). This can help reduce social isolation, increase social capital, and create a more vibrant and connected community.

Third, POS can generate economic benefits such as tourism, business opportunities, and increasing property values (Irwin, 2002). Homebuyers are generally willing to pay more for homes with better views of greenness and openness

(Mak et al., 2010). Finally, POS can contribute to environmental sustainability, biodiversity, and urban climate (Feyisa et al., 2014). Urban greening of parks can help cool towns and cities and mitigate the urban heat island (UHI) effects (Bowler et al., 2010). Strategically located parks and streets can offer a positive thermal effect year-round (Yokohari et al., 2001). POS can also improve air quality (Lewis, 1996). Green vegetation absorbs harmful air pollutants such as carbon dioxide, sulfur dioxide, and nitrogen dioxide and releases oxygen through the process of photosynthesis. Trees and plants in POS help trap dust and other small particles and block their circulation in the air. Different types of green infrastructure can provide a wide range of essential "regulated ecosystem service" to urban settlements (Shao & Kim, 2022). In sum, POS is multifunctional to cities, as they offer many benefits not only to the environment, but also to economic, social, and cultural aspects.

Environmental Equity and Justice

POS is an environmental asset that contributes to the health, social well-being, and economic development of urban populations. However, it is also clear that not all people in cities can benefit equally from this asset. Some disadvantaged communities in urban areas often reside in neighborhoods with inadequate infrastructure, poor living quality, and limited access to green spaces. The mere existence of POS is not enough to ensure their accessibility and enjoyment by the urban population. The spatial location of POS is critical as it determines their proximity to residential areas, transportation networks, and other community facilities. POS that are located far away from residential areas may discourage their use and limit their potential benefits. High development pressure under rapid urbanization often pushes POS away from the core urban areas.

The design and management of POS are also important factors that influence their use and enjoyment. Well-designed POS that cater to the needs and preferences of diverse user groups are more likely to attract visitors and promote social interaction. Effective management practices such as regular maintenance, security, and programming of events also enhance the positive perception of POS. This perception is shaped by cultural, social, and economic factors. Negative perceptions such as safety concerns, lack of amenities, and inadequate maintenance may discourage the use of POS and limit their potential benefits. It is essential to involve the community in the planning and management of POS to ensure their openness, enjoyment, and sustainability. Therefore, environmental justice is not merely about distributive justice, which addresses the availability, accessibility, and attractiveness of POS, it is also about participatory justice and recognition justice, which are concerned about community engagement, equal participation, and rights in decision-making (Kronenberg et al., 2020).

In recent years, the potential role of POS in mitigating adverse urban climate and enhancing public health is increasingly recognized by academia and policymakers (Wu & Ren, 2019; UN-Habitat, 2022). Families of different socioeconomic statuses or ethnicities are found to be facing varying degrees of exposure to environmental hazards (Schweitzer & Stephenson, 2007; Uejio et al., 2011; Agyeman et al., 2016). People living in more densely populated urban areas are exposed to a greater UHI effect and a higher heat-related health risk (Besancenot, 2002). A government study showed that the dense urban cores with a lower amount of POS tended to have much higher heat stress compared with other areas in Hong Kong (Planning Department, 2011). Thus, the lack of supply of POS can exacerbate both undesirable microclimatic effects and negative health outcomes. An uneven distribution of environmental damage can perpetuate socioeconomic inequities when more POS are concentrated in affluent neighborhoods, leading to a further divide between the haves and have-nots. Ultimately, everyone suffers, because poor climate and diseases have no boundaries.

Urban Planning Challenges

Urban planning is both a cause of the above urban inequalities, as well as a solution to environmental injustice (Da Silva et al., 2023). Urban planning is a complex political process that concerns the spatial ordering of activities and involves the allocation of land for various purposes in cities. Urban planners aim to ensure that the available land is utilized in a way that meets the objectives of efficiency, sustainability, equity, safety, and affordability, among others. However, these goals may conflict with each other, and urban planners have to exercise good judgment in their decisions. In many circumstances, POS to be provided by the government tend to have a lower priority in urban land allocation (Tang, 2017). This is because they do not generate direct economic values and are often costly to develop and maintain. Although these spaces provide essential social and environmental benefits to cities, they are often overlooked or sacrificed for other land uses in the planning process (Tang & Wong, 2008).

Poorly managed POS can become trapped in a vicious cycle of poorer maintenance and performance. The deficiency of public investment in a disadvantaged and powerless community can lead to low-quality green infrastructure, limited amenities, and a lack of safety, which in turn reduce the appeal of the POS in its neighborhood. These problems of undersupplied, overused, poorly maintained, and unsafe POS can create a negative feedback loop that further discourages the residents, inside or outside the area, from actively using the spaces. This can lead to less demand or even desertion of POS in these areas, further reducing investment and maintenance to upkeep the functions of POS.

However, increased investment in POS is not necessarily a way out of this impasse, as it can lead to green gentrification (Anguelovski et al., 2022). Public programs to enhance the quantity and quality of POS in a neighborhood can lead to an increased influx of market capital in redeveloping the area. The creation or improvement of POS may generate a facelift of the neighborhood and attract wealthier residents who can afford higher property values. Redevelopment of the old property can result in the displacement of the existing households.

On the other hand, as the demand for housing and commercial spaces increases, landlords may also raise rents, making it unaffordable for low-income households to remain in the area. Therefore, a well-intended POS upgrading strategy may lead to undesirable consequences that sustain rather than reverse environmental inequality.

In professional practice, urban planners often prioritize the recreation and amenities functions of POS, while downplaying its potential role in influencing urban climate. In terms of key performance indicators, the former is easier to quantify than the latter. Take Hong Kong as an example. The planning authority currently defines "open space," as follows:

> any land with the minimum of building structure which has been reserved for either passive or active recreation and provides major or minor recreational facilities, which may be of local or district significance, which is for the use and enjoyment of the general public ... [and it] ... includes park and garden, playground/playing field, promenade, pavilion, sitting out area, pedestrian area and bathing beach.

The current planning standards in Hong Kong aim to allocate at least two square meters of open space per person within a planning area in the town plan. This is incredibly low by international standards. Even worse, it is notoriously unachievable in some dense urban districts. The use of POS by foreign domestic helpers is a notable feature in Hong Kong. On weekdays, they offer help to local families, while on Sundays they usually occupy public spaces and use them throughout the day for leisure (Lai et al., 2021). Urban planners also have guidelines for the location, design, connectivity, facility provision, and greening of POS, but they are not mandatory (Planning Department, 2015). It is clear that all of these planning standards and guidelines focus on the supply and use of POS as a recreational facility for the population, rather than as an environmental asset that helps mitigate damage due to climate change. The lack of a more holistic approach may fail to recognize the multiple benefits that POS can provide to all users in the city.

The division of responsibilities in the government bureaucracy is another factor that can hinder the supply of quality POS. In Hong Kong, for instance, urban planners are only responsible for reserving and allocating land in the town plans for POS, and they have no authority or resources to engage in detailed design, implementation, and management (Tang, 2017). The development, upgrading, and administration of POS are executed by several other government departments separately responsible for land grants, architectural design, construction, and leisure facility management, respectively. A divergence in work priorities, inadequate coordination, and poor communication between departments can lead to delays and a mismatch between the planned and actual outcomes of POS development. Each of these departments needs to work together, but they do not necessarily share the same understanding about the role and functions of POS.

Urban Planning Solutions

Despite these shortcomings, urban planning can also help achieve environmental justice. The goal is to ensure that all residents have equal access to quality POS and desirable environmental assets. A more equitable distribution of accessible, well-connected, and quality POS can lead to a fairer and more equal spread of all benefits. While the recreational and social benefits of POS are important, they can also play a crucial role in mitigating the impacts of climate change. This can be achieved through careful planning, urban design, and management of the urban areas, as well as policies that prioritize the needs of underprivileged communities to create healthier, more sustainable, and more equitable urban communities.

Cities around the world are mainly using two urban planning approaches to mitigate urban heating: (1) altering the land surface cover and (2) modifying urban density, land use, and urban form (Lee & Kim, 2022). Under the first approach, green and blue infrastructure including vegetation, greenery, street trees, and water bodies are found to be useful in reducing heat stress in urban areas (Kurn et al., 1994; Shashua-Bar & Hoffman, 2000; Mullaney et al., 2015; Huang et al., 2016; Turner-Skoff & Cavender, 2019; Wu et al., 2022). Landscaping, green roofs, vertical greening, and shading are among several feasible environmental improvements, especially in underserved neighborhoods (Figure 7.1a). In some dense districts of Hong Kong, urban nullahs or open stormwater drains are being transformed into water parks that serve as new recreational areas and cooling spots for the community. Nonetheless, the cooling effect of blue infrastructure and water features is a knowledge gap in urban heat mitigation and adaptation research (He et al., 2023).

The second approach seeks to alter urban density and morphology in cities. Urban geometry and building density are known to have a strong influence on urban microclimate and intra-urban air temperature (Oke, 1987, 1988). High building volume increases thermal load, worsens air ventilation, and slows down nighttime cooling (Weber & Kuttler, 2003; Hu et al., 2016). Understanding the effects of urban morphology helps in developing appropriate greening strategies against heat waves (Mora et al., 2017). In Japan, for instance, one planning strategy is to use the underground spaces to accommodate some urban uses and "negative" spaces such as transport, communication, energy, and waste disposal, so that more ground areas could be reserved for green POS and essential living space for residents (Nishi et al., 1990; Nishida & Uchiyama, 1993).

Comprehensive renewal of older, dense neighborhoods provides another opportunity to change urban morphology, combat urban decay, and improve the urban environment. Aged buildings that are dilapidated and functionally obsolete are being removed and replaced by newer, greener buildings that meet modern energy and environmental standards. In Hong Kong, large-scale urban renewal projects implemented by the Urban Renewal Authority (URA) normally involve consolidating narrow streets and turning them into much-needed

Figure 7.1 Public spaces in Hong Kong. Examples include (a) landscaping and rooftop greening in a dense urban environment, (b) a street turned into public open space in a comprehensive urban renewal project, (c) public open space provided at the doorstep within a public housing estate, and (d) development of public open spaces underneath a flyover in Kowloon East. Credit: Bo-sin Tang.

at-grade POS within the project areas (Figure 7.1b). This approach helps make up for the shortfalls of POS in the districts, provides more green and open spaces for pedestrians, cyclists, and visitors, improves air ventilation, and enhances the quality of the living and street environment. To preserve the social networks in the existing neighborhood, URA can offer in situ rehousing and move-back arrangements as parts of its compensation packages to affected owners and businesses respectively.

Integrated neighborhood planning and development is another strategy to address the inequality in POS distribution. Poor households in Hong Kong are offered affordable public housing apartments developed by the government. These public housing estates provide a sufficient amount of POS and community facilities at the doorstep (Figure 7.1c). This approach ensures that the public housing residents, especially the elderly and children, have easy walkable access to green spaces and other amenities that promote their health and well-being. In some dense urban areas where ground space is lacking, urban planners may require new private development to provide POS to be located on the podium level of buildings. According to government requirements, these spaces must be freely open to the public and may serve as passive recreational areas for people who do not live on private property. Public open space in private development (POSPD), as this is called in Hong Kong, is a planning gain provided by developers in private development projects. This approach ensures that new developments include public spaces, which is much needed in some districts to serve the community.

Environmental injustice can also be addressed by unconventional approaches. Tactical urbanism, for example, relies on low-cost, temporary measures to transform public spaces. This approach provides quick-fix, immediate benefits to underserved communities. Examples of tactical urbanism include pop-up parks, street art, and mobile markets. Place-making and community engagement are other ways to develop and manage POS. These projects involve collaborations between urban planners and local stakeholders to co-create public spaces, events, and activities that meet community needs. In Hong Kong, Energizing Kowloon East is an example and is famous for its success in transforming the disused spaces under the highways into attractive and vibrant POS for all kinds of community activities (Figure 7.1d).

Lastly, "intergenerational play space" is now a popular design philosophy of POS in Hong Kong. This approach promotes social inclusion, facilitates the use of public spaces by people of all ages, and ensures that public space is accessible to all members of the community. The new POS design in Prosperous Garden, a public housing project of over 1,500 apartments planned and managed by the Hong Kong Housing Society in an old urban district, was one of the pioneering attempts to implement this philosophy. Redesign of the Kowloon Park Fitness Trail into intergenerational play space was another project initiated by a local university in collaboration with government departments and the community. All of these approaches require urban planners to step out of their comfort zones and explore alternative ways to plan, design, and manage urban spaces.

Conclusion

POS contribute to cities in many ways. They offer essential opportunities for physical activity, recreation, social interaction, and cultural events while mitigating the negative impacts of climate change and protecting public health. The unequal distribution of POS is often a reflection of the underlying inequality in society. In many cities, the underserved neighborhoods, typically in low-income areas, have limited access to POS, while the affluent families have an abundance of it. This inequality is sometimes exacerbated by urban planning policies that prioritize economic returns over community needs, and recreation functions over environmental benefits. But the unequal outcomes can also be caused by green gentrification, insufficient public investment, and ineffective government bureaucracy.

Urban planning, however, can also ameliorate this inequality by implementing strategies that prioritize the needs of underserved communities, recognize the multiple values of POS, and facilitate their development, management, and use. Environmental justice can be addressed through investment in blue and green infrastructure, comprehensive urban renewal, integrated neighborhood planning and development in public housing, and POSPD, to provide the much-needed public spaces in the right places. Alternative interventions such as tactical urbanism can be particularly effective in neighborhoods where there is a lack of permanent POS. By means of place-making, community engagement, and embracing the design philosophy of "intergenerational play space," urban planners can create POS that are inclusive to all ages, well-used, and well-loved by the community.

While POS generates multiple benefits, urban planners should also be aware of their possible trade-offs. For instance, overemphasizing the environmental benefits of POS, such as heavy greening and dense vegetation, may result in a disservice to social and economic well-being, such as obstruction of view corridors, blockage of wind flow, poor thermal comfort, high energy consumption, and increases in crime and labor requirements for maintenance (Choi et al., 2021; Probst et al., 2022). Research on green infrastructure pays little attention to social and economic aspects and ignores economic evaluation (Ying et al., 2022). Urban planners need to have a holistic view regarding the impacts of POS. In this way, we can create more sustainable, livable, and inclusive cities that benefit the environment, the economy, and society.

Acknowledgments

The work published in this chapter was funded by the Theme-based Research of the Research Grants Council of the Hong Kong Special Administrative Region (Project No. T-22-504/21-R).

References

Agyeman, J., Schlosberg, D., Craven, L., & Matthews, C. (2016). Trends and directions in environmental justice: From inequity to everyday life, community, and just sustainabilities. In *Annual Review of Environment and Resources* (Vol. 41, pp. 321–340). Annual Reviews Inc. https://doi.org/10.1146/annurev-environ-110615-090052

Anguelovski, I., Connolly, J. J. T., Cole, H., Garcia-Lamarca, M., Triguero-Mas, M., Baró, F., Martin, N., Conesa, D., Shokry, G., del Pulgar, C. P., Ramos, L. A., Matheney, A., Gallez, E., Oscilowicz, E., Máñez, J. L., Sarzo, B., Beltrán, M. A. & Minaya, J. M. (2022). Green gentrification in European and North American cities. *Nature Communications*, *13*(1), 3816–3816. https://doi.org/10.1038/s41467-022-31572-1

Besancenot, J.-P. (2002). Vagues de chaleur et mortalité dans les grandes agglomérations urbaines. *Environnement, Risques & Santé*, 1(4), 229–240.

Bowler, D. E., Buyung-Ali, L., Knight, T. M., & Pullin, A. S. (2010). Urban greening to cool towns and cities: A systematic review of the empirical evidence. *Landscape and Urban Planning*, *97*(3), 147–155. https://doi.org/10.1016/j.landurbplan.2010.05.006

Browning, M., Rigolon, A., McAnirlin, O., & Yoon, H. (2022). Where greenspace matters most: A systematic review of urbanicity, greenspace, and physical health. *Landscape and Urban Planning*, *217*, 13. https://doi.org/10.1016/j.landurbplan.2021.104233

Carmona, M. (2010). Contemporary public space, part two: Classification. *Journal of Urban Design*, 15(2), 157–173. https://doi.org/10.1080/13574801003638111

Choi, C., Berry, P., & Smith, A. (2021). The climate benefits, co-benefits, and trade-offs of green infrastructure: A systematic literature review. *Journal of Environmental Management*, *291*, 16. https://doi.org/10.1016/j.jenvman.2021.112583

da Silva, R. G. P., Lima, C. L., & Saito, C. H. (2023). Urban green spaces and social vulnerability in Brazilian metropolitan regions: Towards environmental justice. *Land Use Policy*, 129, 106638. https://doi.org/10.1016/j.landusepol.2023.106638

Dzhambov, A. M., Lercher, P., Browning, M. H. E. M., Stoyanov, D., Petrova, N., Novakov, S., & Dimitrova, D. D. (2021). Does greenery experienced indoors and outdoors provide an escape and support mental health during the COVID-19 quarantine? *Environmental Research*, 196. https://doi.org/10.1016/j.envres.2020.110420

Farkas, J. Z., Hoyk, E., de Morais, M. B., & Csomos, G. (2023). A systematic review of urban green space research over the last 30 years: A bibliometric analysis. *Heliyon*, *9*(2), 14. https://doi.org/10.1016/j.heliyon.2023.e13406

Feyisa, G. L., Dons, K., & Meilby, H. (2014). Efficiency of parks in mitigating urban heat island effect: An example from Addis Ababa. *Landscape and Urban Planning*, 123, 87–95. https://doi.org/10.1016/j.landurbplan.2013.12.008

He, B. J., Wang, W., Sharifi, A., & Liu, X. (2023). Progress, knowledge gap and future directions of urban heat mitigation and adaptation research through a bibliometric review of history and evolution. *Energy and Buildings*, *287*, 33. https://doi.org/10.1016/j.enbuild.2023.112976

Hu, Y., White, M., & Ding, W. (2016). An urban form experiment on urban heat island effect in high density area. *Procedia Engineering*, 169, 166–174. https://doi.org/10.1016/j.proeng.2016.10.020

Huang, J., Zhou, C., Zhuo, Y., Xu, L., & Jiang, Y. (2016). Outdoor thermal environments and activities in open space: An experiment study in humid subtropical climates. *Building and Environment*, 103, 238–249. https://doi.org/10.1016/j.buildenv.2016.03.029

Irwin, E. G. (2002). The effects of open space on residential property values, *Land Economics*, 78 (4), 465–480.

Kronenberg, J., Haase, A., Łaszkiewicz, E., Antal, A., Baravikova, A., Biernacka, M., Dushkova, D., Filčak, R., Haase, D., Ignatieva, M., Khmara, Y., Niţă, M. R., &

Onose, D. A. (2020). Environmental justice in the context of urban green space availability, accessibility, and attractiveness in postsocialist cities. *Cities*, 106. https://doi.org/10.1016/j.cities.2020.102862

Kurn, D. M., Bretz, S. E., Akbari, H., Laboratory, L. B., & Huang, B. (1994). The Potential for Reducing Urban Air Temperatures and Energy Consumption Through Vegetative Cooling. https://digital.library.unt.edu/ark:/67531/metadc1403828/

Lai, L. W. C., Ho, D. C. W., Chau, K. W., Yu, E. Y. T., Lam, C. L. K., Leung, N. T. H., & Davies, S. N. G. (2021). Property rights & the perceived health contribution of public open space in Hong Kong. *Land Use Policy*, 107. https://doi.org/10.1016/j.landusepol.2021.105496

Lee, S., & Kim, D. (2022). Multidisciplinary understanding of the urban heating problem and mitigation: A conceptual framework for urban planning. *International Journal of Environmental Research and Public Health*, *19*(16), 15. https://doi.org/10.3390/ijerph191610249

Lewis, P. H. (1996). *Tomorrow by Design: A Regional Design Process for Sustainability*. New York: J. Wiley, [1996] ©1996. https://search.library.wisc.edu/catalog/999783072602121

Li, J. J., Mao, Y., Ouyang, J. Y., & Zheng, S. N. (2022). A review of urban microclimate research based on CiteSpace and VOSviewer analysis. *International Journal of Environmental Research and Public Health*, *19*(8), 15. https://doi.org/10.3390/ijerph19084741

Mak, S., Choy, L., & Ho, W. (2010). Quantile regression estimates of Hong Kong real estate prices. *Urban Studies*, 47(11), 2461–2472. https://doi.org/10.1177/0042098009359032

Mora, C., Dousset, B., Caldwell, I., Powell, F., Geronimo, R., Bielecki, C., Counsell, C., Dietrich, B., Johnston, E., Louis, L., Lucas, M., Mckenzie, M., Shea, A., Tseng, H., Giambelluca, T., Leon, L., Hawkins, E., & Trauernicht, C. (2017). Global risk of deadly heat. *Nature Climate Change*, 7. https://doi.org/10.1038/nclimate3322

Mukhamedjanov, A., Kidokoro, T., Seta, F., & Yang, Y. (2021). Reshaping the concept of transit-oriented development in response to public space overheating near the transit nodes of Tokyo. *Cities*, 116. https://doi.org/10.1016/j.cities.2021.103240

Mullaney, J., Lucke, T., & Trueman, S. J. (2015). A review of benefits and challenges in growing street trees in paved urban environments. *Landscape and Urban Planning*, 134, 157–166. https://doi.org/10.1016/j.landurbplan.2014.10.013

Nishi, J., Kamo, F., & Ozawa, K. (1990). Rational use of urban underground space for surface and subsurface activities in Japan. *Tunnelling and Underground Space Technology*, 5(1/2), 23–31.

Nishida, Y., & Uchiyama, N. (1993). Japan's use of underground space in urban development and redevelopment. *Tunnelling and Underground Space Technology*, 8(1), 41–45.

Oke, T. R. (1987). *Boundary Layer Climates* (2nd ed.). London: Methuen & Co.

Oke, T. R. (1988). The urban energy balance. *Progress in Physical Geography: Earth and Environment*, 12(4), 471–508. https://doi.org/10.1177/030913338801200401

Planning Department. (2011). *Urban Climatic Map and Standards for Wind Environment-Feasibility Study: Stakeholders Engagement Digest*. Hong Kong: School of Architecture, The Chinese University of Hong Kong.

Planning Department. (2015). *Hong Kong Planning Standards and Guidelines: Chapter 4 Recreation, Open Space and Greening*. Hong Kong: Government Printer.

Probst, N., Bach, P. M., Cook, L. M., Maurer, M., & Leitao, J. P. (2022). Blue Green Systems for urban heat mitigation: Mechanisms, effectiveness and research directions. *Blue-Green Systems*, *4*(2), 348–376. https://doi.org/10.2166/bgs.2022.028

Schweitzer, L., & Stephenson, M. (2007). Right answers, wrong questions: Environmental justice as urban research. *Urban Studies*, 44(2), 319–337. https://doi.org/10.1080/00420980601074961

Shao, H. M., & Kim, G. (2022). A comprehensive review of different types of green infrastructure to mitigate urban heat islands: Progress, functions, and benefits. *Land, 11*(10), 22. https://doi.org/10.3390/land11101792

Shashua-Bar, L., & Hoffman, M. E. (2000). Vegetation as a climatic component in the design of an urban street An empirical model for predicting the cooling effect of urban green areas with trees. In *Energy and Buildings* (Vol. 31). www.elsevier.comrlocaterenbuild

Tang, B. S. (2017). Is the distribution of public open space in Hong Kong equitable, why not? *Landscape and Urban Planning*, 161, 80–89. https://doi.org/10.1016/j.landurbplan.2017.01.004

Tang, B. S., & Wong, S. W. (2008). A longitudinal study of open space zoning and development in Hong Kong. *Landscape and Urban Planning*, 87(4), 258–268. https://doi.org/10.1016/j.landurbplan.2008.06.009

Tang, B. S. and Wong, S. W. (2021). Soul of the city: Public space and urban planning in Hong Kong, in Mitrasinovic, M. and Jachna, T. (eds.) *The Emerging Public Realm of the Greater Bay Area: Approaches to Public Space in a Chinese Megaregion*. New York: Routledge, pp. 127–135.

Turner-Skoff, J. B., & Cavender, N. (2019). The benefits of trees for livable and sustainable communities. *Plants, People, Planet* 1, 4, 323–335). https://doi.org/10.1002/ppp3.39

Uejio, C. K., Wilhelmi, O. V., Golden, J. S., Mills, D. M., Gulino, S. P., & Samenow, J. P. (2011). Intra-urban societal vulnerability to extreme heat: The role of heat exposure and the built environment, socioeconomics, and neighborhood stability. *Health & Place*, 17(2), 498–507. https://doi.org/10.1016/j.healthplace.2010.12.005

UN-Habitat. (2022). *World Cities Report 2022: Envisaging the Future of Cities*. Nairobi, Kenya: United Nations Human Settlements Programme.

Wan, C., Shen, G. Q., & Choi, S. (2021). Underlying relationships between public urban green spaces and social cohesion: A systematic literature review. In *City, Culture and Society* (Vol. 24). Elsevier Ltd. https://doi.org/10.1016/j.ccs.2021.100383

Weber, S., & Kuttler, W. (2003). Analysis of nocturnal cold-air dynamics and -quality as transported via an urban ventilation path. *Gefahrstoffe Reinhaltung der Luft*, 63, 381–386.

Wu, Z., Man, W., & Ren, Y. (2022). Influence of tree coverage and micro-topography on the thermal environment within and beyond a green space. *Agricultural and Forest Meteorology*, 316. https://doi.org/10.1016/j.agrformet.2022.108846

Wu, Z. F., & Ren, Y. (2019). A bibliometric review of past trends and future prospects in urban heat island research from 1990 to 2017. *Environmental Reviews*, *27*(2), 241–251. https://doi.org/10.1139/er-2018-0029

Ying, J., Zhang, X. J., Zhang, Y. Q., & Bilan, S. (2022). Green infrastructure: Systematic literature review. *Economic Research-Ekonomska Istrazivanja*, *35*(1), 343–366. https://doi.org/10.1080/1331677x.2021.1893202

Yokohari, M., Brown, R. D., Kato, Y., & Yamamoto, S. (2001). The cooling effect of paddy fields on summertime air temperature in residential Tokyo, Japan. *Landscape and Urban Planning*, 53(1–4), 17–27. https://doi.org/10.1016/S0169-2046(00)00123-7

8 Addressing Environmental Quality Inequities

The Role of Information and Collective Action

Sandra Aguilar-Gomez

Introduction

Air and water pollution kills millions of people every year worldwide (Fuller et al., 2022) and, through its impact on human development and economic productivity, hinders the ability of millions more to reach their highest potential (Aguilar-Gomez et al., 2022). Recent studies have highlighted the uneven exposure of economically disadvantaged individuals and people of color to several sources of pollution.[1] Over the past decade, the scope of environmental justice (EJ) research has significantly expanded, marked by advances in data collection and methodologies (Cain et al., 2024). This body of work, mostly focused on the United States, has sought not only to descriptively identify existing disparities but also to investigate their underlying drivers.

A central debate on the drivers of environmental inequities – termed the "sorting vs. siting" debate – discusses whether they are driven mainly by vulnerable populations moving into areas near polluting facilities or by such facilities being sited in proximity to vulnerable communities. The evidence on this matter remains inconclusive,[2] but its policy implications are profound. If residential choices rather than discriminatory siting are the main driver of environmental inequality, the scope for policy intervention is, in principle, smaller. In a framework where individuals with budget constraints choose between environmental quality and other consumption, differences in the marginal willingness to pay (MWTP) for environmental quality can explain some intra- and international differences in exposure to toxic releases (Greenstone and Jack, 2015).

A critical element in this framework is the availability and equal accessibility of information regarding the distribution and risks of pollution. For MWTP to function effectively in a Coasean framework – where clean air is allocated to those who value it most – certain conditions must be met, including complete information. Unequally distributed awareness about environmental hazards entails a significant violation of this condition and perpetuates disparities in pollution exposure among low-income communities and communities of color, leading to broader societal inequities and an environment–health poverty trap (Yang and Liu, 2018).

DOI: 10.4324/9781032631110-9

The issue of information access in EJ debates has been relatively underexplored. The assumption of no informational asymmetries is often implicit. A second required assumption is that transaction costs are low. In a scenario where one side of the bargaining involves a single agent (the local government or a firm) and the other side consists of one or more groups (the population of the area(s) where the noxious facility or other disamenity is to be sited), the assumption of no transaction costs also implies that local collective action is achievable. In this chapter, I elaborate on the pivotal role of information in catalyzing policy change and fostering both individual and collective actions to mitigate the effects of pollution and combat environmental inequalities.

The first section of this chapter, "Environmental Quality Information: Relevance for Individual Behavior, Collective Action, and Policy Change", gives an overview of the recent work on this topic, with an emphasis on causal evidence. Next, in "Sources of Air Pollution Information", I delve into the current landscape of air pollution data. Accessibility of pollution data is crucial for not only the general public but also researchers. I posit that issues of international EJ and disparities in access to environmental quality in developing countries have been underexamined. A dearth of data limits researchers' ability to study these issues and restricts communities' ability to make informed choices and collectively advocate for policy changes. Citizen science initiatives involving the general population in monitoring environmental quality have become innovative data sources for tracking pollution. A "Citizen Science" section presents a discussion of select advancements in integrating environmental science with collective action.

Despite the scant research outside the United States and Europe, significant effort has been exerted to present evidence from other contexts in this chapter. Throughout the review, I examine how the information problem inherent to EJ debates is more profound for developing countries. Figure 8.1 presents the geographical coverage of the empirical papers reviewed in this chapter. While a large portion of the evidence here still comes from North America and Europe, I engage with the growing, high-quality literature on the rest of the world. Finally, although water, soil, and indoor air pollution are crucial issues sporadically touched upon in this review, I focus mostly on outdoor air pollution.

Environmental Quality Information: Relevance for Individual Behavior, Collective Action, and Policy Change

The detrimental effects of pollution on health, productivity, and human capital have been widely documented (Aguilar-Gomez et al., 2022). A growing body of evidence from developing nations finds diverging results on some of these outcomes (Currie et al., 2014). Diverging regional and international impacts reveal potential nonlinearities in the health dose–response function with respect to pollution. Importantly, they might also reveal varying costs of avoidance behaviors, among which accurately assessing pollution levels and damages might be fundamental (Arceo et al., 2016). Individuals with limited

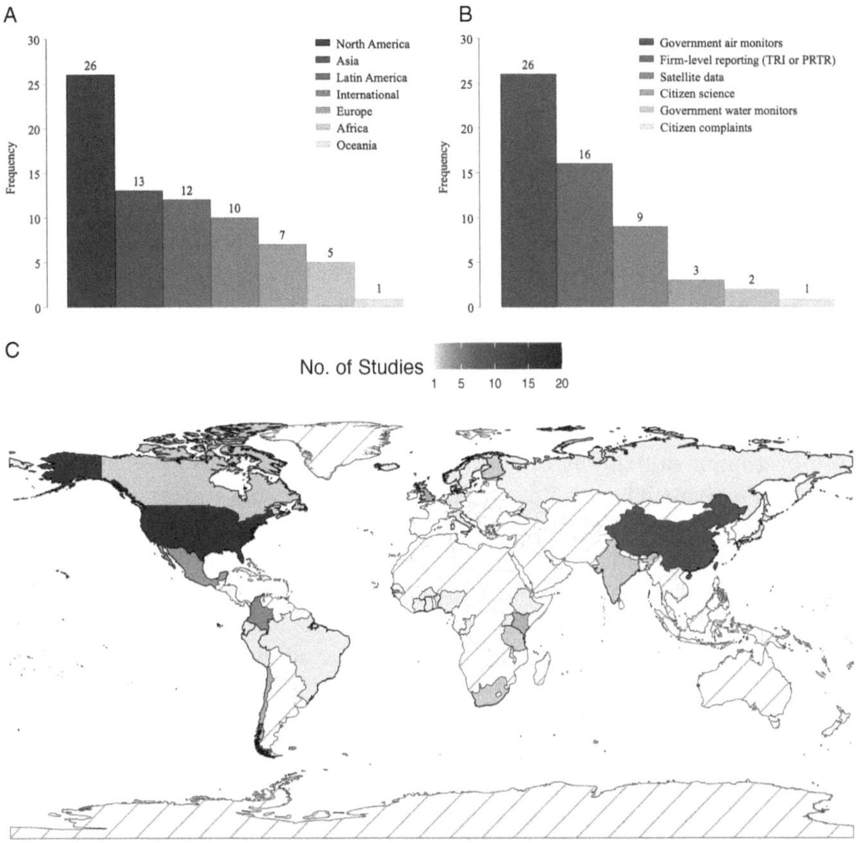

Figure 8.1 Geographic coverage of papers in this review: (a) displays the frequency of each region in the studied papers; (b) displays the frequency of the pollution data sources in the papers that analyze data; and (c) maps the geographical distribution of studies that cover less than 15 countries. Literature reviews and theoretical papers were also excluded. In total, 35 countries for 62 papers were included (countries can appear in more than one paper). Credit: Sandra Aguilar-Gomez.

information may not fully account for the impact of health on income when optimizing utility (Greenstone and Jack, 2015). This is a crucial EJ issue since the uneven distribution of this information between regions and populations has yielded inequality in health gains from pollution control measures (Wang et al., 2023).

Information and Individual Behavior

As a response to minimizing the adverse effects of pollution exposure, avoidance behavior has been a critical subject of study (Zivin and Neidell, 2009; Zivin et al., 2011). When successfully implemented, avoidance serves as an adaptive strategy that can substantially shift the empirically observed relationship between pollution and health outcomes (Neidell, 2009). Such behavior encompasses actions ranging from the use of protective technologies (Zhang and Mu, 2018; Ito and Zhang, 2020) to changes in residential or travel decisions (Chen et al., 2021), and its effectiveness heavily relies on access to accurate pollution information (Aguilar-Gomez et al., 2022).

The link between information failures and disproportionate pollution exposure in low-income communities and communities of color is vastly understudied, as noted by Hausman and Stolper (2021). The authors propose a model where inequality arises when households sort according to known pollution and other disamenities, which are positively correlated with hidden pollution. Crucially, in their framework, even uniformly imperfect information can cause disproportionate pollution exposure and welfare loss among low-income households.

Governments often rely on information dissemination to foster behavioral changes that benefit health. Information can be especially critical in developing countries with severe air pollution. Pollution information tends to be scarcer in these regions. China is a recent exception, as a country with high pollution levels that created significant public awareness through the "war on pollution" while significantly improving state capacity to monitor this phenomenon. Its air quality monitoring program improved public access to pollution information and led to behavioral changes. Barwick et al. (2019) show that these changes significantly mitigated the mortality impact of air pollution, with health benefits outweighing the costs. Zhang and Mu (2018) find that urban residents in China respond to air pollution spikes by increasing their purchases of facemasks, which are effective protective devices.

Sufficient state capacity to produce rapidly available air quality measurements is crucial for alerting the public and prompting action during severe air pollution episodes. While the effectiveness of these programs is debated (Pinder et al., 2019), there is evidence supporting their impact. For instance, studies conducted in Chile (Mullins and Bharadwaj, 2015), Canada (Chen et al., 2018), and Hong Kong (Mason et al., 2019) show that air quality alerts reduce morbidity.

WTP for Environmental Quality

In a simple framework where households have good information on pollution and its health impacts, income strongly predicts MWTP for environmental quality. For poor households, other, more acutely needed goods take precedence. This parameter is often approximated through revealed preference estimates based on agents' responses to environmental quality, such as residential location choices and expenditures on protective devices. Estimates of willingness to pay (WTP) matter: they measure the public's willingness to adopt policies that reduce air pollution and, thus, the political viability of such policies. However, the estimated MWTP for environmental quality may be distorted by market failures and behavioral biases more prevalent in information-poor settings (Greenstone and Jack, 2015).

In this line of work, Ito and Zhang (2020) use data from air purifier purchases to show that heterogeneity in WTP depends not only on household income but also on access to air quality information. To explore this point, the authors exploit a plausibly exogenous information shock: a sudden disclosure by the U.S. Embassy in Beijing in January 2013. Strikingly, their MWTP parameter increases by over 130% after this information shock. Similarly, Tu et al. (2020) find that the environmental documentary "Under the Dome" increased WTP for air quality by 24.9%.

In a study also situated in China but in this case exploring government policy, Wang and Zhang (2023) leverage the implementation of China's air quality information system and analyze high-frequency data on mask purchases. Their findings illustrate that enhanced access to air quality information led to behavioral changes, generating a WTP of approximately 8.6 billion USD annually for a 1 $\mu g/m^3$ reduction in fine particulate matter ($PM_{2.5}$) exposure. This substantial WTP suggests that numerous costly abatement policies could be cost-effective.

It is worth noting that estimates of pollution concerns are not uniformly correlated with income in the literature. For instance, Lin et al. (2021) find that, in the United States, minority groups express greater concern about air and water pollution than their white counterparts, even after socioeconomic status and actual pollution exposure are controlled for.

Information and Collective Action

Collective action is fundamental for resolving common pool resource problems. Local environmental quality, especially air pollution, is an example of a large-scale common pool resource. Environmental protection has long been regarded as the responsibility of the state or regional government. However, collaborative approaches to environmental management are emerging as a relevant complement to purely state-sponsored solutions.

Collective environmental literacy and an understanding of ecosystem services and their dependence on coordination often enhance collaboration. Credibility, relevance, and adaptability within local contexts are vital for information to improve coordination among stakeholders (Opdam et al., 2016; Ardoin et al., 2023). In a particularly interesting example, Magombeyi et al. (2008) use a river basin game to share knowledge among different water users in the Olifants River basin in South Africa. By improving shared understanding of the areas and families affected by lack of access to clean water, the tool facilitated negotiations between upstream and downstream water users that improved collective water management.

The information and communication technology (ICT) revolution presents the promise of improved access to information on collective issues at stake, lower coordination costs, and development of social capital. Focusing on sustainable management of Arctic tourism, Bystrowska et al. (2017) highlight that ICTs improve collective action by increasing visibility, creating proximity, and introducing new management practices. However, in some contexts, some modes of local collective action can be hindered by ICT access. Perilla (2019) finds that internet access in Colombia reduced political engagement, evident in lower voter turnout, decreased candidate votes, and increased blank votes. Similarly, in a literature review, Sobbrio (2014) highlights that increasing the supply of news media can negatively impact electoral participation under a crowding-out effect. Successful ICT-enabled coordination requires high social capital, subject to social norms and existing relationships (Bystrowska et al., 2017).

Information, Government Responses, and Policy Change

Accurate and detailed environmental monitoring is not only a tool for policymakers in enforcing regulations but also a catalyst for policy improvements. The public's understanding of the relevance of pollution and climate change abatement policies is crucial to their success. Greenstone and Hanna (2014) contrast the success of air pollution abatement policies in India between 1986 and 2007 with the failure of water policies. High citizen engagement led India's Supreme Court to promptly promulgate and enforce pollution regulations. The court was responding to high media coverage and public awareness of the high levels and hazardous consequences of polluted air, in contrast to the low public concern with water pollution. The air policies studied were more effective in cities with higher literacy and greater newspaper attention to the problems of air pollution.

Significant environmental reforms rely heavily on broad public support for policy change. To maintain this support, it is crucial not only to adequately compensate those who might be adversely affected but also to ensure that adequate information on the benefits of reform reaches the most disadvantaged communities. Dechezleprêtre et al. (2022) study public support for climate policies in 20 countries and find that people's perception of the policies' effectiveness and impact on lower-income groups is crucial. Their findings show the need for clear communication about the benefits and workings of these policies. Policymakers must demonstrate how climate policies can lead to positive outcomes, especially for the most vulnerable groups.

The research by Buntaine et al. (2022) on China presents a complementary perspective on the importance of public access to pollution information. They find that public engagement, facilitated through social media, significantly influenced regulatory actions. Similarly, Araujo et al. (2024) study the impact of an increase in public attention on local environmental action, focusing on the viral 2019 Brazilian Amazon forest fires. The research reveals that increased public scrutiny led to a 22% reduction in fires in Brazil, preventing the emission of 24.8 million tons of CO_2. Both studies highlight the influential role of public attention in driving action on critical environmental issues.

Information empowers communities, influences regulatory bodies, and drives technological innovations. Technological progress is a game-changer under these dynamics. Its impact is threefold. First, the decline in costs of public environmental monitoring systems has propelled their massive rollout in several developing countries and previously underserved areas in developed nations. Second, the spatial and temporal granularity of satellite data has massively increased, lowering the costs of reacting swiftly to pollution events. Third, low-cost monitor deployment, often linked to citizen science initiatives, has enabled the public scrutiny discussed in this chapter. In the next section, I discuss the scope and impact of diverse pollution information sources.

Sources of Air Pollution Information

Information not only empowers individuals to make informed decisions to protect their health but also plays a pivotal role in the enforcement of environmental policies and the identification of disparities in environmental degradation. What information sources do citizens have to monitor pollution? Are these equally available across countries and socioeconomic groups? In this section, I review various sources of pollution information that are instrumental in facilitating avoidance behaviors and pollution mitigation efforts. Focusing on air pollution, I describe potential biases in access to these information sources, which contribute to unequal exposure to environmental hazards.

Ground Monitoring Stations

Air pollution monitoring stations are a crucial input for government surveillance and academic research. Half of the empirical studies discussed in this chapter employ monitoring stations as their main data source. However, they are still scarce in many regions and often suboptimally placed.

$PM_{2.5}$ is a key pollutant to monitor given that its size allows it to enter the bloodstream, with significant short- and long-run effects (Aguilar-Gomez et al., 2022). There is a significant gap in air quality monitoring in low- and middle-income countries (LMICs): currently, at least 1,900 LMIC cities lack $PM_{2.5}$ monitoring capabilities. In contrast, in Europe and North America, nearly every city has several ambient air quality $PM_{2.5}$ monitors. Larger cities often have a dozen or more, averaging approximately one monitor for every 100,000–600,000 residents. In stark contrast, across urban areas in Africa, the average is one monitor per 4.5 million residents (Pinder et al., 2019).

At the subnational level, monitoring station placement is uneven and often not representative: socioeconomic and political considerations can lead to biased placement, further exacerbating the challenge of effectively responding to environmental hazards. Misplacement of pollution monitoring stations has been documented, at least in the United States. Grainger and Schreiber (2019) find that local regulators take demographic characteristics into account when designing ambient air pollution networks. Regulators avoid pollution hotspots when installing monitors, and avoidance is even stronger when the area is poor.

The timing of network functioning can also be endogenous to geographical and temporal disparities in air quality. Mu et al. (2021) find across the United

States evidence of strategic timing of monitor shutdowns to coincide with air quality alerts. The monitors showing these patterns are predominantly in federal nonattainment areas that could face costly penalties for violations of U.S. Clean Air Act standards. Zou (2021) employs satellite data to show that, in the United States, air quality significantly worsens on days without monitoring (once every six days) because of short-term suppression of pollution on monitored days.

Satellite Data

Researchers have made strides in filling the gaps in sparse ground-based data by spatial interpolation (Janssen et al., 2008; Mejía et al., 2023). However, these methods perform poorly in capturing anomalous values and require a large historical dataset (Mejía et al., 2023).

Satellite remote sensing has emerged as a promising source of complementary or alternative data on air quality. Multiple space-based measurements and global chemical transport models are combined to create annual and monthly ground-level $PM_{2.5}$ estimates with resolutions as low as 1.1 km^2 (e.g., Van Donkelaar et al., 2021). Nitrogen oxides (NOx), important for climate change, groundwater quality, and linked to cancer risk, are also increasingly monitored from space. Measurements of NO_2 in the European satellite data product Sentinel-5P, available since 2018, have been found to be consistent with ground-based measurements (Ialongo et al., 2020) but offer global coverage.

Satellite data can function as a tool to prioritize locations for accurate ground-level air quality monitoring implementation. Guevara Luna et al. (2018) illustrate this argument for PM in Colombia, a country where few cities possess advanced air quality monitoring networks, leaving many areas lacking reliable data. Using NASA's MODIS satellite and surface measurements, the study identifies areas with elevated aerosol optical depth (AOD) levels that lack ground-level monitoring in three major cities.

Satellite data can provide valuable insights into environmental inequalities in regions with limited ground-based monitoring. Aguilar-Gomez et al. (2024) utilize satellite-derived $PM_{2.5}$ estimates and geolocalized household surveys to show that racialization plays a significant role in determining exposure to worsening air quality in Colombia.

Pollutant Release and Transfer Registries

A pollutant release and transfer register (PRTR) is a nationwide database tracking releases of potentially hazardous chemicals and pollutants into the air, water, and soil and their off-site transfer for treatment or disposal. The first such program, the Toxic Release Inventory (TRI), initiated in 1986, marked the beginning of systematic public disclosure of chemical emissions by companies in the United States and spurred improvements in environmental practices (EPA, 2023). The TRI paved the way for PRTRs in other countries.

While PRTRs are implemented mainly in wealthier OECD countries, efforts to widen adoption continue.

PRTRs in Developed Countries

Households' responses to the TRI have been documented mostly through behavior in housing markets. Initial evidence for the United States reflected little housing market response to the TRI (Bui and Mayer, 2003). However, more recent evidence finds a significant impact. Mastromonaco (2015) finds that TRI guidelines strengthening reporting requirements decrease housing prices within approximately 1 mile of existing firms newly subject to reporting by 11%. Currie et al. (2015) uncover that the opening of TRI-subject plants also results in an 11% decline in housing values within 0.5 miles, equivalent to a loss of approximately $4.25 million for affected households, with impacts concentrated in relatively disadvantaged communities. Finally, Hibiki and Managi (2011) examine the Japanese PRTR's impact on the rental housing market. Residents recognize carcinogenic risks within 0.5 km–1.0 km of facilities, leading to a 0.0007% rent reduction for every 1% increase in risk.

Firms faced with PRTR requirements often underreport emissions (de Marchi and Hamilton, 2006). However, they also seek to enhance efficiency to lower emissions in certain situations. For instance, Manello (2017) finds that, in Italy and Germany, chemical companies eventually invested in improving efficiency after initially struggling with compliance with the European Pollution Release and Transfer Register (E-PRTR).

Environmental compliance can be costly for firms, significantly altering their location decisions (List et al., 2003). These location costs can have significant environmental and climate justice implications at a larger scale that can differ from their local implications for cleanup. Tang (2015) finds evidence of emissions shifting from the United Station to poorer countries with weaker environmental protections after a chemical is included in the TRI registry. Tanaka et al. (2022) find a similar pollution haven phenomenon in Mexico after the tightening of U.S. lead standards in 2009. In response to regulation, firms relocated their battery recycling operations to Mexico, where significant repercussions for infant health occurred. The pollution haven phenomenon is an archetypal example of international environmental injustice and a challenge to be acknowledged when countries engage in unilateral efforts to reduce toxic emissions.

The fact that individual, identifiable firms must comply with PRTR reporting requirements makes them a powerful tool for identifying the sources of toxic exposure and enforcing improvements. However, the requirements' complexity has called into question whether these tools, in themselves, can foster collective and political action (Bui and Mayer, 2003). In this discussion, Bae et al. (2010) argue that, while the TRI aims to reduce community information costs, raw data are ineffective mainly because they necessitate sophisticated interpretation. State-led initiatives that refine and analyze TRI data offer communities a clearer understanding of health risks, enabling them to advocate for reduced toxic releases.

PRTRs in Developing Countries

Mexico and Chile are the only LMICs with fully developed PRTRS. However, data inconsistencies in these countries and other OECD nations have been frequently highlighted, indicating a substantial need for enhancements in national enforcement. There is also a pressing need for international standardization regarding the range of pollutants reported, industries required to report, methods for measuring emissions, and reporting thresholds (Mardones and Mena, 2020; Milani, 2017). Research on such registries' impacts on pollution abatement comparable to the evidence reviewed above is virtually nonexistent. A notable exception is Chakraborti (2022), who examines reporting behavior under Mexico's PRTR of 3,436 major polluters in response to environmental inspections. Inspections and fines improve reporting of toxic releases. When followed by fines, an inspection also leads to a 50% reduction in annual releases.

An emerging strand of work has employed these data sources to assess the consequences of, and disparities in, toxic exposure. Campos and Mardones (2023) use PRTR data from Chile to link copper production to health and agriculture effects. Chakraborti and Shimshack (2022) use Mexican PRTR data on toxic water pollution discharges from over 1600 industrial facilities. The inventory quality of this dataset, theoretically covering the universe of relevant emitters, allowed them to link this information to urban block–level outcomes. They find that industrial facilities tend to pollute more in marginalized neighborhoods and, that environmental disparities in urban Mexico are associated with collective action, community pressure, and Coasean bargaining.

Citizen Science

Traditionally, air and water quality monitoring has depended on expert-driven siting of instruments to gauge city-wide average concentrations of chemicals and particles. Citizen science alternatives are emerging as a valuable resource for filling information gaps and addressing inequalities associated with pollution exposure and impact. With advancing technology and decreasing costs, there is a growing trend in both high-income countries and Asian, African, and Latin American LMICs to deploy a vast number of lower-cost sensors with collaboration from the general public (Pinder et al., 2019; Mahajan et al., 2022).

Citizen science programs have served different purposes depending on the context where they are deployed. Sometimes, they are designed as educational initiatives to foster interest in science, environmental knowledge attitudes and practices (Hadjichambi et al., 2023), and social capital around environmental issues (Van Brussel and Huyse, 2019). Ideally, citizen science can also be integrated into public management, improving evidence-based policymaking (Ngo et al., 2017; deSouza et al., 2017).

Citizen science initiatives designed with an EJ perspective recognize that certain sectors are disproportionately affected by environmental issues and seek to include such sectors as program beneficiaries (Mahajan et al., 2022). They can be a tool for generating smart cities that are citizen-centric and environmentally and socially sustainable (Mahajan et al., 2021).

The survey by Elias et al. (2023) provides data on the landscape of citizen science projects in Africa. Out of 53 projects, the majority were initiated for research advancement (27 projects), with the remainder related to public education (23), influencing policy (11), and collecting data from end users (10). However, the study finds limited integration of citizen science data as an input for government agencies in the region. A challenge faced by many participatory science initiatives is the difficulty of balancing high-quality data, deep citizen engagement, and significant policy influence, as noted by Van Brussel and Huyse (2019).

Citizen science can also provide solutions to complement and enhance the enormous potential of satellite imagery and traditional monitoring sensors. Most of these projects offer data at higher spatial resolution than traditional monitoring networks and higher spatiotemporal resolution than satellite data for the covered areas. An interesting example of this synergy is Picture Pile, an initiative that has led to the collection of more than 11 million classifications of images. Picture Pile has been used in many domains, including detecting deforestation, conducting post-disaster assessments, monitoring poverty, and identifying marine litter (Fraisl et al., 2022).

In a successful and inclusive citizen science project, the engagement plan is defined from the outset and maintained continuously from project initiation to completion. Antwerp's CurieuzeNeuzen involved over 2000 citizens in the deployment of passive air quality sensors. The data and feedback from this project were utilized for academic research and contributed to policy debates on air quality (Van Brussel and Huyse, 2019). In sum, citizen sensing projects should acknowledge the concerns of affected communities and shape their research questions accordingly (Mahajan et al., 2022).

Discussion

Air pollution remains responsible for approximately 9 million deaths per year, corresponding to one in six deaths worldwide (Fuller et al., 2022). In Africa alone, each year without clean indoor and outdoor air leads to the premature deaths of over 700,000 people (Roy, 2016). Significant disparities in pollution exposure have been found in the United States (Cain et al., 2024), and more recently, in other contexts.

While the link between information failures and disproportionate pollution exposure in low-income communities and communities of color is vastly understudied, I present an overview of the robust evidence of the crucial importance of information access for individual protective behavior, local collective action, and regulatory and enforcement improvements.

I discuss the role of different sources of environmental quality information, focusing on air pollution. When designing ambient air pollution networks, authorities should aim to obtain representative data and deploy ground monitoring stations with appropriate coverage of disadvantaged and especially polluted areas. Satellite data should be used to fill gaps in the ground network data, and participatory approaches can be leveraged to obtain granular data and engage local populations.

National registries of toxic emissions and facilities are increasingly used to raise public awareness of potential health hazards in local areas. Such registries would represent the most detailed source of toxic release information for governments, and with enough enforcement and the right dissemination scheme, they can help reduce noxious emissions that poison local environments. An unintended consequence, however, may be the offshoring of production to less regulated countries. I highlight the importance of concerted transboundary to global toxic abatement efforts for preventing international environmental injustice.

International disparities in pollution levels and monitoring capacities are highlighted throughout this chapter. 52% of the papers in this review refer to LMICs, despite the intended emphasis on such regions. These disparities underscore the need to expand air quality monitoring and research in the Global South.

Acknowledgements

Daniela Vlasak-Gonzalez provided outstanding research assistance.

Notes

1 Reviews of this literature include Agyeman et al. (2016); Cushing et al. (2018); Banzhaf et al. (2019); Baker et al. (2020) and Cain et al. (2024).

2 Rüttenauer and Best (2021) find limited evidence supporting the selective siting hypothesis but stronger evidence for selective migration in Germany. Conversely, Cushing et al. (2018) find a correlation between historic red-lining practices and the siting of fossil fuel power plants causing contemporary disparities in air pollution in the United States. Adding another dimension to this debate is the "steering" hypothesis, which argues that housing market discrimination limits the residential choices available to minorities (Christensen et al., 2022).

References

Aguilar-Gomez, S., Dwyer, H., Graff Zivin, J. and Neidell, M. (2022). This is air: The "nonhealth" effects of air pollution. *Annual Review of Resource Economics*, 14, 403–425.

Aguilar-Gomez, S., Cárdenas, J. C., and Salas Díaz, R. (2024). Environmental justice beyond race: Skin tone and exposure to air pollution. *Documento CEDE*, (7), 1–51. http://doi.org/10.2139/ssrn.4742122

Agyeman, J., Schlosberg, D., Craven, L. and Matthews, C. (2016). Trends and directions in environmental justice: From inequity to everyday life, community, and just sustainabilities. *Annual Review of Environment and Resources*, 41, 321–340.

Araujo, R., Costa, F. J. M. and Garg, T. (2024). Going viral: Public attention and environmental action in the Amazon. *Journal of the Association of Environmental and Resource Economists*, 11 (6), 1605–1627.

Arceo, E., Hanna, R. and Oliva, P. (2016). Does the effect of pollution on infant mortality differ between developing and developed countries? Evidence from Mexico City. *The Economic Journal*, 126 (591), 257–280.

Ardoin, N. M., Bowers, A. W. and Wheaton, M. (2023). Leveraging collective action and environmental literacy to address complex sustainability challenges. *Ambio*, 52 (1), 30–44.

Bae, H., Wilcoxen, P. and Popp, D. (2010). Information disclosure policy: Do state data processing efforts help more than the information disclosure itself? *Journal of Policy Analysis and Management*, 29 (1), 163–182.

Baker, K. R., Amend, M., Penn, S., Bankert, J., Simon, H., Chan, E., Fann, N., Zawacki, M., Davidson, K. and Roman, H. (2020). A database for evaluating the InMAP, APEEP, and EASIUR reduced complexity air-quality modeling tools. *Data in Brief*, 28, 104886.

Banzhaf, H. S., Ma, L. and Timmins, C. (2019). Environmental justice: Establishing causal relationships. *Annual Review of Resource Economics*, 11, 377–398.

Barwick, P. J., Li, S., Lin, L. and Zou, E. (2019). *From fog to smog: The value of pollution information*. National Bureau of Economic Research.

Bui, L. T. and Mayer, C. J. (2003). Regulation and capitalization of environmental amenities: Evidence from the Toxic Release Inventory in Massachusetts. *Review of Economics and Statistics*, 85 (3), 693–708.

Buntaine, M., Greenstone, M., He, G., Liu, M., Wang, S. and Zhang, B. (2022). *Does the squeaky wheel get more grease? The direct and indirect effects of citizen participation on environmental governance in China*. National Bureau of Economic Research.

Bystrowska, M., Wigger, K. and Liggett, D. (2017). The use of information and communication technology (ICT) in managing high arctic tourism sites: A collective action perspective. *Resources*, 6 (3), 33.

Cain, L., Hernandez-Cortes, D., Timmins, C., and Weber, P. (2024). Recent findings and methodologies in economics research in environmental justice. *Review of Environmental Economics and Policy*, 18(1), 116–142.

Campos, L. and Mardones, C. (2023). An economic evaluation of the health and agricultural damages caused by copper mining in Chile. *Latin American Research Review*, 58 (2), 342–367.

Chakraborti, L. (2022). Environmental deterrence in developing countries: Evidence from enforcement actions and fines under the toxics monitoring program of Mexico. *Environment and Development Economics*, 27 (6), 511–532.

Chakraborti, L. and Shimshack, J. P. (2022). Environmental disparities in urban Mexico: Evidence from toxic water pollution. *Resource and Energy Economics*, 67, 101281.

Chen, H., Li, Q., Kaufman, J. S., Wang, J., Copes, R., Su, Y. and Benmarhnia, T. (2018). Effect of air quality alerts on human health: A regression discontinuity analysis in Toronto, Canada. *The Lancet Planetary Health*, 2 (1), e19–e26.

Chen, S., Chen, Y., Lei, Z. and Tan-Soo, J.-S. (2021). Chasing clean air: Pollution-induced travels in China. *Journal of the Association of Environmental and Resource Economists*, 8 (1), 59–89.

Christensen, P., Sarmiento-Barbieri, I. and Timmins, C. (2022). Housing discrimination and the toxics exposure gap in the United States: Evidence from the rental market. *Review of Economics and Statistics*, 104 (4), 807–818.

Currie, J., Zivin, J. G., Mullins, J. and Neidell, M. (2014). What do we know about short-and long-term effects of early-life exposure to pollution? *Annual Review of Resource Economics* ,6 (1), 217–247.

Currie, J., Davis, L., Greenstone, M. and Walker, R. (2015). Environmental health risks and housing values: Evidence from 1,600 toxic plant openings and closings. *American Economic Review*, 105 (2), 678–709.

Cushing, L., Blaustein-Rejto, D., Wander, M., Pastor, M., Sadd, J., Zhu, A. and Morello-Frosch, R. (2018). Carbon trading, co-pollutants, and environmental equity: Evidence from California's cap-and-trade program (2011–2015). *PLoS Medicine*, 15 (7), e1002604.

Dechezleprêtre, A., Fabre, A., Kruse, T., Planterose, B., Chico, A. S. and Stantcheva, S. (2022). *Fighting climate change: International attitudes toward climate policies*. National Bureau of Economic Research.

deSouza, P., Nthusi, V., Klopp, J. M., Shaw, B. E., Ho, W. O., Saffell, J., Jones, R. and Ratti, C. (2017). A Nairobi experiment in using low-cost air quality monitors. *Clean Air Journal*, 27 (2), 12.

Elias, P., Shonowo, A., de Sherbinin, A., Hultquist, C., Danielsen, F., Cooper, C., Mondardini, M., Faustman, E., Browser, A., Minster, J.-B. et al. (2023). Mapping the landscape of citizen science in Africa: Assessing its potential contributions to sustainable development goals 6 and 11 on access to clean water and sanitation and sustainable cities. *Citizen Science: Theory and Practice*, 8 (1), 33.

Fraisl, D., Campbell, J., See, L., Sturn, T., MacFeely, S., Bowser, A., Campbell, J., Moorthy, I., Danylo, O., McCallum, I. and Fritz, S. (2022). Demonstrating the potential of Picture Pile as a citizen science tool for SDG monitoring. *Environmental Science & Policy*, 128, 81–93.

Fuller, R., Landrigan, P. J., Balakrishnan, K., Bathan, G., Bose-O'Reilly, S., Brauer, M., Caravanos, J., Chiles, T., Cohen, A., Corra, L. et al. (2022). Pollution and health: A progress update. *The Lancet Planetary Health*, 6 (6), e535–e547.

Grainger, C. and Schreiber, A. (2019). Discrimination in ambient air pollution monitoring? In *AEA Papers and Proceedings, American Economic Association 2014 Broadway, Suite 305, Nashville, TN 37203*, vol. 109, pp. 277–282.

Greenstone, M. and Hanna, R. (2014). Environmental regulations, air and water pollution, and infant mortality in India. *American Economic Review*, 104 (10), 3038–3072.

Greenstone, M. and Jack, B. K. (2015). Envirodevonomics: A research agenda for an emerging field. *Journal of Economic Literature*, 53 (1), 5–42.

Guevara Luna, M. A., Guevara Luna, F. A., Espinosa, J. F. M. and Cerón, L. C. B. (2018). Spatial and temporal assessment of particulate matter using AOD data from MODIS and surface measurements in the ambient air of Colombia. *Asian Journal of Atmospheric Environment*, 12 (2), 165–177.

Hadjichambi, D., Hadjichambis, A. C., Adamou, A. and Georgiou, Y. (2023). *A systematic literature review of K-12 environmental citizen science (CS) initiatives: Unveiling the CS pedagogical and participatory aspects contributing to students environmental citizenship*. Educational Research Review, p. 100525.

Hausman, C. and Stolper, S. (2021). Inequality, information failures, and air pollution. *Journal of Environmental Economics and Management*, 110, 102552.

Hibiki, A. and Managi, S. (2011). Does the housing market respond to information disclosure?: Effects of toxicity indices in Japan. *Journal of Environmental Management*, 92 (1), 165–171.

Ialongo, I., Virta, H., Eskes, H., Hovila, J. and Douros, J. (2020). Comparison of tropomi/sentinel-5 precursor no 2 observations with ground-based measurements in Helsinki. *Atmospheric Measurement Techniques*, 13 (1), 205–218.

Ito, K. and Zhang, S. (2020). Willingness to pay for clean air: Evidence from air purifier markets in China. *Journal of Political Economy*, 128 (5), 1627–1672.

Janssen, S., Dumont, G., Fierens, F. and Mensink, C. (2008). Spatial interpolation of air pollution measurements using CORINE land cover data. *Atmospheric Environment*, 42 (20), 4884–4903.

Lin, R., Ma, L. and Phan, T. (2021). Race and environmental worries. Available at SSRN 3918520.

List, J. A., McHone, W. W. and Millimet, D. L. (2003). Effects of air quality regulation on the destination choice of relocating plants. *Oxford Economic Papers*, 55 (4), 657–678.

Magombeyi, M. S., Rollin, D. and Lankford, B. (2008). The river basin game as a tool for collective water management at community level in South Africa. *Physics and Chemistry of the Earth, Parts A/B/C*, 33 (8–13), 873–880.

Mahajan, S., Luo, C.-H., Wu, D.-Y. and Chen, L.-J. (2021). From do-it-yourself (DIY) to do-it-together (DIT): Reflections on designing a citizen-driven air quality monitoring framework in Taiwan. *Sustainable Cities and Society*, 66, 102628.

Mahajan, S., Chung, M.-K., Martinez, J., Olaya, Y., Helbing, D. and Chen, L.-J. (2022). Translating citizen-generated air quality data into evidence for shaping policy. *Humanities and Social Sciences Communications*, 9 (1), 1–18.

Manello, A. (2017). Productivity growth, environmental regulation and win–win opportunities: The case of chemical industry in Italy and Germany. *European Journal of Operational Research*, 262 (2), 733–743.

de Marchi, S. and Hamilton, J. T. (2006). Assessing the accuracy of self-reported data: An evaluation of the Toxics Release Inventory. *Journal of Risk and Uncertainty*, 32 (1), 57–76.

Mardones, C. and Mena, C. (2020). Effects of the internalization of the social cost of global and local air pollutants in Chile. *Energy Policy*, 147, 111875.

Mason, T. G., Schooling, C. M., Chan, K. P. and Tian, L. (2019). An evaluation of the air quality health index program on respiratory diseases in Hong Kong: An interrupted time series analysis. *Atmospheric Environment*, 211, 151–158.

Mastromonaco, R. (2015). Do environmental right-to-know laws affect markets? Capitalization of information in the Toxic Release Inventory. *Journal of Environmental Economics and Management*, 71, 54–70.

Mejía, D., Alvarez, H., Zalakeviciute, R., Macancela, D., Sanchez, C. and Bonilla, S. (2023). Sentinel satellite data monitoring of air pollutants with interpolation methods in Guayaquil, Ecuador. *Remote Sensing Applications: Society and Environment*, 31, 100990.

Milani, S. (2017). The impact of environmental policy stringency on industrial R&D conditional on pollution intensity and relocation costs. *Environmental and Resource Economics*, 68, 595–620.

Mu, Y., Rubin, E. A. and Zou, E. (2021). *What's missing in environmental (self-) monitoring: Evidence from strategic shutdowns of pollution monitors*. National Bureau of Economic Research.

Mullins, J. and Bharadwaj, P. (2015). Effects of short-term measures to curb air pollution: Evidence from Santiago, Chile. *American Journal of Agricultural Economics*, 97 (4), 1107–1134.

Neidell, M. (2009). Information, avoidance behavior, and health the effect of ozone on asthma hospitalizations. *Journal of Human Resources*, 44 (2), 450–478.

Ngo, N., Kokoyo, S. and Klopp, J. (2017). Why participation matters for air quality studies: Risk perceptions, understandings of air pollution and mobilization in a poor neighborhood in Nairobi, Kenya. *Public Health*, 142, 177–185.

Opdam, P., Coninx, I., Dewulf, A., Steingröver, E., Vos, C. and van der Wal, M. (2016). Does information on landscape benefits influence collective action in landscape governance? *Current Opinion in Environmental Sustainability*, 18, 107–114.

Perilla, S. (2019). Internet access and electoral participation in Colombia: Voter apathy at the polls. Available at SSRN 3759775.

Pinder, R. W., Klopp, J. M., Kleiman, G., Hagler, G. S., Awe, Y. and Terry, S. (2019). Opportunities and challenges for filling the air quality data gap in low-and middle-income countries. *Atmospheric Environment*, 215, 116794.

Roy, R. (2016). *The cost of air pollution in Africa*. OECD Development Centre, Working Paper No. 333. https://www.oecd-ilibrary.org/docserver/5jlqzq77x6f8-en.pdf?expires=1728404160&id=id&accname=guest&checksum=151EFF846DDE5B4715E73A866514C9F4

Rüttenauer, T. and Best, H. (2021). Environmental inequality and residential sorting in Germany: A spatial time-series analysis of the demographic consequences of industrial sites. *Demography*, 58 (6), 2243–2263.

Sobbrio, F. (2014). The political economy of news media: Theory, evidence and open issues. In *A handbook of alternative theories of public economics*, 278. https://www.elgaronline.com/edcollchap/edcoll/9781781004708/9781781004708.00021.xml

Tanaka, S., Teshima, K. and Verhoogen, E. (2022). North-south displacement effects of environmental regulation: The case of battery recycling. *American Economic Review: Insights*, 4 (3), 271–288.

Tang, J. P. (2015). Pollution havens and the trade in toxic chemicals: Evidence from US trade flows. *Ecological Economics*, 112, 150–160.

Tu, M., Zhang, B., Xu, J. and Lu, F. (2020). Mass media, information and demand for environmental quality: Evidence from the "Under the Dome". *Journal of Development Economics*, 143, 102402.

US Environmental Protection Agency (EPA), (2023). *Toxics Release Inventory (TRI) program*. United States Environmental Protection Agency.

Van Brussel, S. and Huyse, H. (2019). Citizen science on speed? Realising the triple objective of scientific rigour, policy influence and deep citizen engagement in a large-scale citizen science project on ambient air quality in Antwerp. *Journal of Environmental Planning and Management*, 62 (3), 534–551.

Van Donkelaar, A., Hammer, M. S., Bindle, L., Brauer, M., Brook, J. R., Garay, M. J., Hsu, N. C., Kalashnikova, O. V., Kahn, R. A., Lee, C. et al. (2021). Monthly global estimates of fine particulate matter and their uncertainty. *Environmental Science & Technology*, *55*(22), 15287–15300.

Wang, Z. and Zhang, J. (2023). The value of information disclosure: Evidence from mask consumption in China. *Journal of Environmental Economics and Management*, 122, 102865.

Wang, Z., Liu, J., Wang, B., Zhang, B. and Deng, N. (2023). Health benefits from risk information of air pollution in China. *Scientific Reports*, 13 (1), 15432.

Yang, T. and Liu, W. (2018). Does air pollution affect public health and health inequality? empirical evidence from China. *Journal of Cleaner Production*, 203, 43–52.

Zhang, J. and Mu, Q. (2018). Air pollution and defensive expenditures: Evidence from particulate-filtering facemasks. *Journal of Environmental Economics and Management*, 92, 517–536.

Zivin, J. G. and Neidell, M. (2009). Days of haze: Environmental information disclosure and intertemporal avoidance behavior. *Journal of Environmental Economics and Management*, 58 (2), 119–128.

Zivin, J. G., Neidell, M. and Schlenker, W. (2011). Water quality violations and avoidance behavior: Evidence from bottled water consumption. *American Economic Review*, 101 (3), 448–453.

Zou, E. Y. (2021). Unwatched pollution: The effect of intermittent monitoring on air quality. *American Economic Review*, 111 (7), 2101–2126.

9 Locating Representational Equity

Intersections between Child-Friendly Cities and Children's Roles in Realizing Environmental Justice

Victoria Derr

Introduction

> What is the point of having green space, if you can't breathe the air outside?

This question, posed by a university student from Bakersfield, California, a community with one of the highest rates of air pollution in the state (OEHHA 2023), highlights an essential shortcoming in equitable built environments. Children and youth around the world consistently identify safe access to natural areas and healthy environments as an important aspect of realizing child-friendly cities, yet these rights are inconsistently realized, primarily along income, race and ethnicity, and class lines (Adams et al. 2019; Forni et al. 2019). These disparities occur between countries of the Global North and Global South, across cities with disparate incomes and wealth, and within neighborhoods and districts of cities themselves. Consistent factors inhibit children's realization of these rights: lack of equity in the distribution and quality of green spaces; non-existent, inaccessible, or unsafe transportation systems; inequitable distribution of factories, freeways, and waste facilities; lack of social safety and sense of belonging; high crime and violence; and discriminatory planning and policy systems that stigmatize youth or fail to include marginalized voices and address inequities at appropriate scales (Adams et al. 2019; Forni et al. 2019; Gülgönen and Corona 2015; Johnston et al. 2020; Méndez 2020; Rigolon, Browning, and Jennings 2018). However, children and youth themselves often identify the explicit ways these injustices manifest in their everyday lives and are able to articulate meaningful pathways for change (Derr, Chawla, and Mintzer 2018).

Child-friendly cities are a potential means to articulate and realize children's rights to more equitable environments. UNICEF conceptualized the Child Friendly Cities Initiative (CFCI) in 1996 as a way to promote children's rights in municipal decision-making and government (UNICEF 2023). The rights-based framework draws from the 1989 United Nations Convention on the Rights of the Child (CRC), which is one of the most widely adopted frameworks for human rights in the world. Both the CRC and CFCI promote rights

DOI: 10.4324/9781032631110-10

to children's health and wellness and the development of their education, capabilities, and participation in decision-making (OHCHR 2023). As of 2023, UNICEF's CFCI has been adopted by more than 3,000 city governments in more than 40 communities worldwide (UNICEF 2023).

The CRC of 1989, on which the CFCI is based, includes 54 articles that delineate children's rights. The Convention articulates multiple provisions for children's rights to freedom of expression, influence over policies and practices, and influence over decision-making that impacts children's access to education, health, and safe environments for play and leisure. Article 24 articulates children's rights to environmental sanitation, nutritious foods, and clean water and air. Article 29 includes "the development of respect for the natural world" as a specific goal of children's education (OHCHR 2023).

Development of children and youth participation in local planning initiatives also has been supported through the United Nations Educational, Scientific and Cultural Organization (UNESCO) Growing Up in Cities program. First developed by Kevin Lynch in 1970 and later revived in the 1990s, the Growing Up in Cities program developed participatory methods with young people that are the foundation for many participatory planning and design projects today (Chawla and van Vliet 2020; Derr, Chawla, and Mintzer 2018). In the 1990s revival, Growing Up in Cities brought the rights-based focus of the CRC of 1989 together with the participatory methods Lynch had developed (Chawla and van Vliet 2020). These practices have since been integrated into multiple spatial scales in low- and high-resourced neighborhoods and cities around the world (Derr, Chawla, and Mintzer 2018). Growing Up Boulder is perhaps the longest-standing partnership that has sought to realize both the rights-based approaches of child-friendly cities along with the participatory practices from Growing Up in Cities since its inception in 2009 (Derr, Chawla, and Mintzer 2018; Mintzer and Chawla 2020).

Frameworks and articulations of rights are insufficient for realizing environmentally just and inclusive spaces unless they also are deliberately and systematically woven into existing policies, plans, and practices. A 2022 review of published literature about child-friendly cities identified consistency in the general indicators used to describe child-friendly environments but that the extent that child-friendly cities are realized often aligns with sociodemographic and environmental injustices that exist both within and across cities of different sociodemographic statuses (Jansson et al. 2022). However, as both environmental justice and child-friendly approaches advance, they provide important examples and lessons for how rights to a healthy environment can be realized even across disparities of income, race, or class.

This chapter provides an overview of UNICEF's CFCI and the recent launch of a CFCI pilot program within the United States. Programs and projects from environmental justice communities that center children's rights and inclusion both serve to highlight ways that cities are approaching equity in the built environment and ways that cities can address disparities to better realize children's rights to a healthy environment.

UNICEF's Child Friendly Cities Initiative

The UNICEF CFCI is a child-rights-based framework to promote the well-being of children. The framework involves bringing together local governments, community partners, and children and youth themselves so that young people have the ability to participate in decisions that affect them. CFCI emphasizes children's participation, safety and inclusion, access to social services, safe living environments, and play and leisure. Environmental justice issues often cut across these goal areas (UNICEF USA personal communication, June 2, 2023).

There is a six-step process to becoming a recognized UNICEF Child Friendly City: (i) formalizing the CFCI through a Memorandum of Understanding; (ii) engaging community stakeholders, including through youth council or other youth leadership structures; (iii) conducting a situation analysis in which local governments assess existing data and identify gaps; (iv) holding intergenerational community conversations to determine priority issues and develop a Local Action Plan; (v) implementing the action plan and demonstrating results for children; and (vi) undergoing an independent evaluation of CFCI implementation. The successful completion of these steps leads to recognition by UNICEF as a Child Friendly City. The CFCI process is designed as an iterative cycle so that government, community, and children and youth stakeholders are embedded into local processes and practices. These strategies allow capacity building among both young people and adult decision-makers. Successful CFCI cities map child-friendly initiatives onto existing strategic plans and resources rather than creating external add-ons or locating practices in a siloed department. A clearly identified coordinator within a lead agency can be housed in any range of departments, such as parks and recreation, public health, education, or community planning, but stakeholders draw from many departments and sectors for project planning and implementation (UNICEF USA personal communication, June 2, 2023).

All CFCI cities must demonstrate how nondiscrimination and inclusion as well as youth participation have been integrated into all phases of the process. Inclusion and participation can be institutionalized through youth councils, a children's budget, an ombudsperson who represents children's perspectives, or child and youth impact assessments (UNICEF USA personal communication, June 2, 2023).

The United States is an important context for piloting CFCI because it is the only United Nations (UN) member nation that has not ratified the UN CRC. Local governments do not have the same level of accountability to enact the CRC-rooted principles of child-friendly cities, but local municipalities may choose to enact these principles independently. In addition, UNICEF launched the CFCI pilot in the United States in 2020, in the middle of a global pandemic, racial reckoning, and climate crisis (UNICEF USA personal communication, June 2, 2023). Approaches to these issues are likely to resonate with communities who experience environmental injustices or who seek to center equity and children's rights in the built environment.

Environmental Justice and the Realization of Children's Rights in the Built Environment

Child-friendly cities are important components of urban planning and development because the majority of the global population lives in urban areas, and nearly a third of urban residents are children. According to their website, UNICEF estimates that by 2050, 70% of the world's children will live in cities, and many of these children will live in slums or substandard living conditions. Cities are thus sites of responsibility and opportunity. Children and youth understand the complexity of cities; they are experts in their lived experiences and they are capable of informing solutions and decision-making at young ages. Decades of work now support this, and stronger integration into urban governance is increasingly recognized as an important aspect of not only realizing children's rights but also of realizing equity in the built environment (Derr, Chawla, and Mintzer 2018).

Prior to the launch of the pilot initiative, UNICEF USA conducted a National Situation Analysis in 2019 which identified that more than 10 million children in the United States were living in neighborhoods without sufficient access to quality education, safe outdoor play spaces, nutritious food, or clean air, and that racial and geographic disparities existed in communities across the country. Youth leadership in central issues, such as the climate crisis or school shootings, combined with global and digital youth organizing, also align with increasing investment in mayors and local governments to drive change. The CFCI provides opportunities to achieve local change alongside young people and promotes opportunities to support some of the cities' most vulnerable groups, including urban Indigenous communities and newly arriving migrant children. UNICEF USA recognizes that inequities in the realization of children's rights are often based on zip code and extend along environmental, social, and racial lines. Equity therefore is at the forefront of local change and putting children's rights into practice (UNICEF USA personal communication, June 2, 2023).

Because issues emerge from the process with children and youth themselves, the CFCI emphasizes the realization of *procedural justice* in specific social and environmental contexts of each city. UNICEF USA encourages cities to incorporate community engagement and child rights awareness-raising opportunities in every stage of the process to build interest and demand for CFCI. In doing so, cities also help children and youth develop agency to advocate for their rights and to build capacity among local leaders to ensure that children's rights are upheld through government processes (UNICEF USA personal communication, August 21, 2023).

Two cities that are participating in the CFCI pilot – Houston, Texas and Boulder, Colorado – also exemplify ways to connect children's rights with ideas of *distributive environmental justice*. In Houston, one way this is achieved is through the adoption of a Children's Outdoor Bill of Rights in April 2023. Established by an intergenerational committee of environmental leaders and youth, the bill articulates 12 rights for equitable nature accessibility, regardless of zip code, race, or income. These rights include provisions for safe and accessible routes to outdoor places, access to nature in school settings, leadership and learning opportunities, and the ability to breathe clean air outdoors (City of Houston 2023). Houston also identified child-responsive emergency planning as a priority issue. The City of Houston, in particular, has been identified as one of the most vulnerable urban centers in the world because of its proximity to the Gulf of Mexico and related risks from more frequent and intense tropical storms, hurricanes, and floods. These storms have caused disproportionate harm to predominantly Black, Hispanic, and low socioeconomic status neighborhoods within the city (Chakraborty, Collins, and Grineski 2019). During the CFCI planning process, a diverse network of children and community members identified the need for child-centered emergency response and improved resource accessibility before, during, and after emergencies. This presented an opportunity for the city to integrate a child-rights lens and child-responsiveness into its disaster and emergency response planning (UNICEF USA personal communication, June 2, 2023).

In Boulder, the city is mapping equity in access to nature and parks for children ages 0–5. The composite analysis will serve to identify high-priority areas for policymaking and future CFCI project initiatives (UNICEF USA personal communication, June 2, 2023). Houston and Boulder's efforts are supported by Cities Connecting Children to Nature, a joint initiative of the National League of Cities and the Children & Nature Network which is designed to address racial disparities in access to nature (NLC 2023). Since 2009, the City of Boulder also has integrated children and youth perspectives into decision-making for parks and open space planning, resilience planning, housing, transportation, eco-healing, and climate action through its Growing Up Boulder program. Drawing on child-friendly frameworks, Growing Up Boulder achieves these goals through community, city government, and school partnerships focused on urban planning needs as well as youth-identified, emergent issues. Children and youth themselves have centered nature as an important focal point across a range of projects, and are often strong advocates for the rights of nature in city planning (Mintzer and Chawla 2020).

Around the world, however, scholars who study child-friendliness have identified nature as an important and underrepresented aspect of children's own conceptualizations of child-friendly cities. In a study of child-friendly environments in Cape Flats, South Africa, for example, children described a "culture of inconsideration" toward nature, in which the natural environment was polluted and degraded. In this study, children "revered" ideals of safe natural spaces for play but reflected that these spaces neither existed within their community nor could they safely access the outdoors due to violence. However, these same children had solutions. Children proposed stronger integration of environmental education into school curricula and community centers where they could safely create access to natural spaces and increase nutrition for children and families with the provision of enclosed community gardens and soup kitchens (Adams et al. 2019). In Mexico City, children similarly described an ideal city as one replete with nature, animals, and clean air and contrasted this with their hostile, unsafe, and unclean environment (Gülgönen and Corona 2015). Notably, while the CFCI framework and CRC address children's rights to learn about and experience healthy environments, they do not explicitly identify equity in children's access to nature as a right.

Realizing Equity and Child-Friendliness, No Matter Where You Are

UNICEF's CFCI emphasizes procedural steps to engage children and youth in decision-making through formal inclusion in municipal government. This can take the form of youth councils, partnerships between city departments and schools, or other means of embedding children and youth participation into urban planning, design, and governance. However, it takes institutional capacity and human resources to realize the "gold standard" of CFCIs in an integrated way (UNICEF USA personal communication, June 2, 2023). While there are clear benefits to doing so, including that children's rights are more likely to be consistently realized and sustained across time, there are many communities who seek to realize these rights outside formal recognition of the CFCI. In some cases, youth themselves have identified compelling reasons for working outside established frameworks, including that youth have viewed youth councils as elitist and nonrepresentative or see them as mechanisms to control youth dissent rather than empower youth in making change (Taft and Gordon 2013). In addition, environmental justice communities and organizations often work outside local governments due to a lack of trust and accountability (Harrison 2019) and because spatial scales of city planning often minimize environmental justice concerns that are experienced at neighborhood scales (Méndez 2020).

Some of the most consistent mechanisms for realizing children's rights in environmental justice contexts have been through partnerships between youth-serving groups or schools, academics, and environmental justice organizations. Some communities also work with designers to implement small-scale projects that identify from a child-centered lens how children's rights can be manifest through the built environment (Derr, Chawla, and Mintzer 2018). Because children and youth often focus on smaller scales – parks, streets, schools, or neighborhoods – their concerns and recommendations also tend to highlight inequities and injustices that may be overlooked otherwise. Like environmental justice issues more broadly, these spatial scales are important for identifying ways to build a more just, equitable, and healthy city.

Three of the most common areas of focus include air quality, ecological restoration and equitable access to nature, and climate justice. Each of these issues intersects with children's ability to realize their rights to a safe and healthy environment. Conversely, failure to systematically address these issues as part of environmentally just and child-friendly cities poses widespread health risks for children (Durrwachter-Erno et al. 2021). The examples below all draw from communities that experience environmental injustices due to discriminatory policies and practices that have led either to disproportionate distribution of environmental harms and/or to lack of access to environmental benefits, primarily along race, class, and income lines.

Air Quality

Turin, Italy is home to the Fiat car industry and an associated high rate of automobile-oriented transportation infrastructure. The working-class city hosts the highest levels of air pollution in Italy, and among the highest in Europe. As part of the Bernard van Leer Foundation's Urban95 Challenge, a team of architects, environmental scientists, urbanists, and planners launched a project to address air pollution and promote a healthier and more playable city. Starting with Urban95's question, "if you could see the city from an elevation of 95 cm – the average height of a healthy 3-year-old child – what would you do differently?", the project examined how "bottom-up" approaches could address socially disadvantaged children's vulnerability to health hazards stemming from poor air quality (Forni et al. 2019, 53). Projects focused on air quality are particularly significant for equity in the built environment for children because children are more exposed to and susceptible to the detrimental effects of pollution, leading to a higher risk of respiratory illness, such as asthma (Forni et al. 2019; Johnston et al. 2020). The project in Turin involved strategies for awareness creation through public meetings and teacher and parent training, knowledge building and participation of children themselves, and monitoring of both indoor and outdoor air quality. The outcomes were used to inform city planning strategies that could reduce environmental harm (Forni et al. 2019, 54–55). This project highlights that even young children are capable of being involved in action research and making recommendations for improving the quality of their city. This is consistent with a range of projects supported by Growing Up Boulder over many years that similarly engage young children in participatory planning (Derr, Chawla, and Mintzer 2018).

The participatory approach applied in Turin is commonly applied in many communities which seek to address environmental health disparities among youth. For example, in Los Angeles County, California, where communities of color are disproportionately situated near industrial facilities and freeways, and where government regulatory monitors do not provide sufficient data resolution for neighborhood or personal scales, a partnership between university academics and youth-based environmental justice organizations examined environmental justice through a suite of methods. The *Day in the Life* project gathered neighborhood-level air quality monitoring data that supplemented city-scale data, generated maps that linked air quality with the daily routines of youth, and generated photographic and video documentation of youth experiences and their daily lives (Johnston et al. 2020). Like the Turin, Italy project, this participatory monitoring activity sought to increase environmental health literacy and build capacity among youth and community members to understand and apply air quality data to inform public health action. In addition to structured training and capacity building, *Day in the Life* also enhanced youth storytelling skills and collaboration between scientists, youth, and environmental justice organizations. Some of the youth shared their research with school officials and with the Los Angeles City Council. These efforts are connected to other programs that demonstrate disproportionate burdens and advocate for more equitable policies and planning (Personal communication, Jill Johnston email, June 16, 2023).

The Historic West End neighborhood of Charlotte, North Carolina brought air quality monitoring and youth engagement together with local actions and policy change to realize environmental justice in an African American neighborhood that has experienced disproportionate amounts of air pollution due to urban heat and discriminatory housing and infrastructure practices (cleanAIREnc 2024; Durrwachter-Erno et al. 2021). In the two-year Clean Air Carolina project, students in five schools in the Historic West End learned about air pollution, applied citizen science and air quality monitoring strategies, and generated maps to show pollution levels and differential exposures. Some participating schools also generated bike monitoring systems. Students learned about air quality codes and zoning and how to protect their health during poor air quality periods. At the schools, children and youth implemented action projects, including an ozone bio-monitoring garden, idle free school policies, and strategies for addressing risks from wildfire smoke. Several neighborhood-scale initiatives have been promoted to shift toward more equitable planning and permitting processes and cleaner construction practices within the neighborhood (cleanAIREnc 2024).

Ecological Restoration

Young people have been engaged as active participants in the ecological restoration of urban landscapes for many decades as part of urban community forestry and community garden initiatives (Campbell and Wiesen 2009). For example, the Urban Resources Initiative was established in Baltimore, Maryland in 1989 and brought to New Haven, Connecticut in 1991 to engage youth from disenfranchised neighborhoods in restoring local ecosystems and establishing local control (URI, n.d.). Principles of participation have been integral to these programs, creating opportunities for historically excluded communities to shape urban ecosystem management. Such approaches often engage young people in community projects that restore, replant, and rehabilitate community spaces and often emphasize social revitalization alongside ecological restoration (Campbell and Wiesen 2009; Krasny 2018). Increasingly, young people themselves are identifying connections between race, class, and power and are working inter-generationally to increase their access to environmental commons. Youth directly link the health of land and the health of people and identify ways they can address toxic air, soil, and water. Youth often bring their own cultural capital to the framing of such engagements, bringing arts and hip-hop together with more traditional stewardship practices (Gallay et al. 2016).

With the increasing impacts of climate change on urban heat, attention also has been placed on the associations between lack of tree canopy and urban heat islands as an environmental injustice stemming from discriminatory redlining policies (Locke et al. 2021). The City of Houston's Children's Outdoor Bill of Rights specifically identifies provisions for shade and climate change mitigation as principles of an equitable outdoor environment (City of Houston 2023). Houston is the fifth and largest city in the United States to implement a Children's Outdoor Bill of Rights (following the cities of Austin, Baltimore, Salt Lake City, and San Francisco) (City of Houston 2023). Only Houston and Baltimore's bill of rights explicitly identify environmental justice within the framework, and only Houston articulates specific mechanisms to address environmental and climate injustices, including youth leadership and participation in stewardship activities as well as safe access to nature (City of Houston 2023).

The United Nations Decade on Ecosystem Restoration, 2021–2030, aims to regenerate healthy ecosystems while also seeking to address poverty alleviation, climate change, and species extinctions. The initiative specifically calls for young people's inclusion, and Challenge 6.1 seeks to embed ecosystem restoration into the environmental education practices of schools and organizations worldwide (UN Decade of Ecosystem Restoration 2021).

Climate Justice

The City of Oakland, California is an example of how youth can be integrated into a planning process in a systematic way, even while working outside formalized child-friendly frameworks. The City of Oakland has a long history of both disenfranchisement as well as civil rights leadership through the Black Panthers, Oakland Community School, and Ella Baker Center for Human Rights which have promoted youth agency and organizing (Hogan 2019). Oakland's approach is informative for strategies to advance equity in children's rights and children's safe access to healthy cities because they intentionally focused on local and embodied knowledge, encouraged broad public participation of its diverse residents, and focused on human scale impacts that would provide health benefits (Méndez 2020, Chapter 7).

Oakland's process directly involved more than 30 community stakeholder groups, including many youth-serving organizations. The resulting 2030 Equitable Climate Action Plan identifies strategies for building reciprocity with local Indigenous communities, addressing local climate change impacts, and building frontline community resilience (City of Oakland 2020). Youth engagement included a solar-powered concert with famous hip-hop artists; a climate adaptation work day that brought more than 350 residents, including many youth, to install gardens and rainwater catchment systems at a local college; and more than 80 high school students who envisioned climate solutions in their homes, schools, and neighborhoods (Méndez 2020, Chapter 7).

Through the Skyline High School's Green Energy Pathway program and Y-PLAN (a youth participatory research initiative at the University of California, Berkeley), more than 100 youth also researched low-carbon mobility and made recommendations for climate actions that made their way to additional workshops, youth commission meetings, and final climate plans (City of Oakland 2020).

Some engagement processes reflect the "Oaklandish" character of Oakland, including the Original Scraper Bike Team, comprised of East Oakland youth who create artistically refurbished bicycles and bring awareness to bikes as a more affordable and greener mode of transportation while also raising awareness about teen violence (The Original Scraper Bike Team 2023). In the context of climate action planning, Scraper Bike youth made recommendations for street murals and other street transformations that would support more active biking and visible crosswalks (City of Oakland 2020). To mitigate climate anxiety among youth, training and job programs that support youth in regenerative agriculture, green infrastructure, ecological restoration, and city greening also were integral parts of the plan (City of Oakland 2020). This trajectory builds on Ella Baker Center's work (Hogan 2019).

Oakland's Equitable Climate Action Plan is notable for the way community voices, including those of youth, were integrated throughout the entire process and embedded into the final plan. The Oakland case study also shows how working for climate justice at a local level is also a means toward achieving strategies for greater equity and inclusion overall.

Conclusion

This chapter highlights ways that children's rights can be realized and embedded into the design and planning of more equitable cities. Many projects represented here reflect principles of local communities working together and integrating child advocates and young people themselves into processes of urban governance, decision-making, and action. Lessons from UNICEF's Child Friendly Cities Initiative and Growing Up Boulder highlight best practices to support more equitable and inclusive cities. Many of these lessons have been articulated in previous publications (e.g., Chawla and van Vliet 2020; Derr, Chawla, and Mintzer 2018). Some of the most salient recommendations for achieving equity in the built environment include:

1 *Recognize that child-friendly city frameworks are about realizing rights over time*. Like environmental justice itself, realizing rights and developing structures that embrace the principles of child rights and child-friendliness take time and strong relationships. Building trust and identifying common goals are essential aspects of the process across levels of government, community groups, and children and youth themselves.
2 *Embed children and youth strategies, policies, and practices into existing structures within a city*. Identify adult champions within existing

organizations or city departments, and formulate broad partnerships that represent diverse interests and sectors of the population of interest, whether at a neighborhood or city scale.

3 *Explicitly identify strategies for achieving equity* with each new process, project, or framework adoption. Seek to develop changes in culture as well as systems. Cultural shifts are often needed to realize child-friendliness within cities, and systemic change is part of realizing equity in the built environment. CFCI's core principles of nondiscrimination and inclusion are essential for environmental justice projects as well. History has shown that without deliberate attention to equity, children and youth of color, from working-class families, or with lower incomes, will not reap equal benefits from municipal-scaled planning. Working at scales that allow disparities to be understood and addressed is also a part of realizing equity.
4 *Identify both processes for participation and products or outcomes that can lead to discernable results*. Process outcomes are important to children and youth in their development of agency and their ability to realize change. Processes are also important to support youth in feeling heard, which is an undervalued aspect of city planning. However, finding ways to authentically integrate children and youth ideas into tangible outcomes is also important. This can be through small-scale projects, such as the ozone biomonitoring gardens in Charlotte, or through the integration of youth interests through Oakland's comprehensive Equitable Climate Action Plan. Participation in itself is not emancipatory or liberating unless it is done well and ultimately leads to results. Identify actions and who will be responsible: city leaders, community organizations or groups, or coalitions of young people themselves.
5 *Create transparent systems for accountability, reflection, and evaluation*. Children, youth, community partners, and city staff all need time to reflect on successes and accomplishments, as well as ways to better advance child-friendly practices or policies. As partnerships grow, create clear and transparent systems for how decisions are made and resources are allocated. This will allow all stakeholders to evaluate whether policies and practices support equity.

When cities decide to draw on the international principles of children's rights and frameworks for participation, they are beginning a cultural shift that can result in more socially and environmentally just spaces that are often healthier and more desirable (Derr, Chawla, and Mintzer 2018, Chapter 1). Several decades of work within this international context has led to a strong body of work with children and youth from around the world, in the Global North and Global South, in highly resourced cities and informal settlements. There are many paths to identifying community strengths and concerns and to promoting change. Young people have shown they can be active participants in realizing change, and child-friendly city frameworks provide both ethical and practical strategies to achieve this. When young people are engaged, they often extend their thinking to people of other ages, abilities, ethnicities, and interests, as well as to other species in nature (Derr, Chawla, and Mintzer 2018, Chapter 11). In many instances, young people navigate ways to change themselves by changing the city, thus articulating in their own contexts their "right to the city" (Gallay et al. 2016, 121). Realizing child friendliness in cities can thus be a way to realize equity in the built environment.

When I worked with Growing Up Boulder from 2012 to 2016, we aspired to receive UNICEF recognition as a Child Friendly City, but at that time we were still envisioning how to realize child-friendliness in the context of Boulder. Instead, we drew from UNICEF's self-evaluation tools, determined how we could center equity and inclusion in local contexts, and experimented with strategies for meaningful child and youth participation through a variety of projects and community partnership structures. In 2023, the time was right, and Boulder's Child Friendly City Initiative was launched as part of the UNICEF USA pilot program, further embedding structures for child-friendliness across city governance and culture. As someone who now works in much smaller communities with few resources and many environmental injustices, it is helpful for me to remember that everyone starts somewhere. The communities in this chapter show aspirational models of participation at a variety of scales as well as ways to identify and articulate the challenges and opportunities specific to a local community context. Systematically realizing children's rights and child-friendly cities is a long-term commitment, but to paraphrase Lao Tzu, it is a journey that begins with a single step.

Acknowledgments

Thank you to Amber Stone, Danielle Goldberg, and the UNICEF USA team for providing detailed information on the Child Friendly Cities Initiative and for reviewing the manuscript, and to Louise Chawla and Mara Mintzer from Boulder, Colorado, for sharing current contexts and resources that informed this manuscript.

References

Adams, Sabirah, Shazly Savahl, Maria Florence, and Kyle Jackson. 2019. "Considering the Natural Environment in the Creation of Child-Friendly Cities: Implications for Children's Subjective Well-Being." *Child Indicators Research* 12 (February): 545–567.

Campbell, Lindsay, and Anne Wiesen, eds. 2009. *Restorative Commons: Creating Health and Well-Being through Urban Landscapes*. General Technical Report, NRS-P-39. Newtown Square, PA: U.S. Department of Agriculture, Forest Service Northern Research Station.

Chakraborty, Jayajit, Timothy W. Collins, and Sara E. Grineski. 2019. "Exploring the Environmental Justice Implications of Hurricane Harvey Flooding in Greater Houston, Texas." *American Journal of Public Health* 109, no. 2 (February): 244–250. https://doi.org/10.2105/AJPH.2018.304846

Chawla, Louise, and Willem van Vliet 2020. "Welcoming young people in urban place-making: Learning from challenges." In *The Routledge Handbook of Designing Public*

Spaces for Young People, edited by Janet Loebach, Sarah Little, Adina Cox, and Patsy Eubanks Owens, 120–134. New York: Routledge.
City of Houston. 2023. "*Children's Outdoor Bill of Rights.*" Mayor's Office of Education. Accessed June 6, 2023. https://www.houstontx.gov/education/cobor.html
City of Oakland. 2020. "Oakland 2030 Equitable Climate Action Plan." Accessed June 7, 2023. https://www.oaklandca.gov/projects/2030ecap
cleanAIREnc. (2024). "Clearing the air: Charlotte's Historic West End." Accessed June 7, 2023 https://cleanairenc.org/
Derr, Victoria, Louise Chawla, and Mara Mintzer. 2018. *Placemaking with Children and Youth: Participatory Practices for Planning Sustainable Communities*. New York: New Village Press.
Durrwachter-Erno, Katie, Gredia Huerta-Montanez, Vivian Nguyen, Aaron A. Levy, Jennifer Mah Lawson, and Vi Thuy Nguyen. 2021. "Informing Policy on Built Environments to Safeguard Children in Environmental Justice Communities: Case Study of Five AAP Climate Advocates." *Journal of Applied Research on Children* 12, no. 1: 1. https://doi.org/10.58464/2155-5834.1451
Forni, Elisabetta, Emanuele Negro, Chiara Carlucci, Angela Nasso, and Mirjam Struppek. 2019. "Actions against Air Pollution in Turin for a Healthy and Playable City." *Cities & Health* 3, no. 1–2 (May): 53–58. https://doi.org/10.1080/23748834.2019.1596045
Gallay, Erin, John Lupinacci, C. Sarmiento, C. Flanagan, and Ethan Lowenstein. 2016. "Youth environmental stewardship and activism for the environmental commons." In *Contemporary Youth Activism: Advancing Social Justice in the United States*, edited by Jerusha Conner and Sonia M. Rosen, 113–134. Westport, Connecticut: Praeger.
Gülgönen, Tuline, and Yolanda Corona. 2015. "Children's Perspectives on their Urban Environment and Their Appropriation of Public Spaces in Mexico City." *Children, Youth and Environments* 25, no. 2: 208–228. https://doi.org/10.7721/chilyoutenvi.25.2.0208
Harrison, Jill. 2019. *From the Inside Out: The Fight for Environmental Justice within Government Agencies*. Cambridge, MA: MIT Press.
Hogan, W. 2019. *On the Freedom Side: How Five Decades of Youth Activists Have Remixed American History*. Chapel Hill, North Carolina: University of North Carolina Press.
Jansson, Märit, Emma Herbert, Alva Zalar, and Maria Johansson. 2022. "Child-Friendly Environments—What, How and by Whom?" *Sustainability* 14, no. 8 (April): 4852. https://doi.org/10.3390/su14084852
Johnston, Jill, Zully Juarez, Sandy Navarro, Ashley Hernandez, and Wendy Gutschow. 2020. "Youth Engaged Participatory Air Monitoring: A 'Day in the Life' in Urban Environmental Justice Communities." *International Journal of Environmental Research and Public Health* 17, no. 1 (December): 93. https://doi.org/10.3390/ijerph17010093
Krasny, Marianne, ed. 2018. *From Grassroots to Global: Broader Impacts of Civic Ecology*. Ithaca, NY: Cornell University Press.
Locke, Dexter, Billy Hall, J. Morgan Grove, Steward Pickett, Laura Ogden, Carissa Aoki, Christopher Boone, and Jarlath PM O'Neil-Dunne. 2021. "Residential Housing Segregation and Urban Tree Canopy in 37 US Cities." *npj Urban Sustainability* 1, no. 1 (March): 15. https://doi.org/10.1038/s42949-021-00022-0
Méndez, Michael. 2020. *Climate Change from the Streets: How Conflict and Collaboration Strengthen the Environmental Justice Movement*. New Haven, CT: Yale University Press.
Mintzer, Mara, and Louise Chawla. 2020. "Turning young people's ideas into action: Learning from long-term partnerships for child and youth participation." *The Routledge Handbook of Designing Public Spaces for Young People*, edited by Janet Loebach, Sarah Little, Adina Cox, and Patsy Eubanks Owens, 135–149. Routledge.
National League of Cities. 2023. "Cities Connecting Children to Nature." Accessed June 5, 2023. https://www.nlc.org/initiative/cities-connecting-children-to-nature/

OEHHA. 2023. CalEnviroScreen 4.0. "California Office of Environmental Health Hazard Assessment". Accessed May 31, 2023. https://oehha.ca.gov/calenviroscreen

Office of the High Commissioner on Human Rights. 2023. "*Convention on the Rights of the Child. General Assembly resolution 44/25. Adopted 20 November 1989.*" United Nations Office of the High Commissioner on Human Rights. Accessed June 1, 2023. https://www.ohchr.org/en/instruments-mechanisms/instruments/convention-rights-child

Rigolon, Alessandro, Matthew Browning, and Viniece Jennings. 2018. "Inequities in the Quality of Urban Park Systems: An Environmental Justice Investigation of Cities in the United States." *Landscape and Urban Planning* 178 (October): 156–169. https://doi.org/10.1016/j.landurbplan.2018.05.026

Taft, Jessica K., and Hava R. Gordon. 2013. "Youth Activists, Youth Councils, and Constrained Democracy." *Education, Citizenship and Social Justice* 8, no. 1: 87–100. https://doi.org/10.1177/1746197913475765

The Original Scraper Bike Team. 2023. "About." Accessed June 8, 2023. https://scraperbiketeam.com

UN Decade on Ecosystem Restoration. 2021. Accessed June 7, 2023. https://www.decadeonrestoration.org/

UNICEF. (2023) "Child Friendly Cities." Accessed June 6, 2023. https://www.childfriendlycities.org/

Urban Resources Initiative (URI). n.d. "About." Accessed June 6, 2023 https://uri.yale.edu/about

10 Disparities in the Obesogenic Built Environment for Children

A Pilot Study in East Baton Rouge Parish, Louisiana

Yutian Zeng, Senlin Chen, Amanda E. Staiano, and Fahui Wang

According to the National Health Statistics Reports, the obesity prevalence in the United States among children and adolescents aged 2–19 years increased from 17% in 2011–2014 to 19.7% in 2017–2020, and the rate was even higher at 20.7% for school-age children aged 6–11 in 2017–2020 (Ogden et al., 2018; Stierman et al., 2021). Childhood obesity prevalence increased during the COVID-19 pandemic, further intensifying this public health problem (Lartey et al., 2023). Obesity contributes to other disease conditions such as hypertension, type 2 diabetes, and respiratory problems (Shaw et al., 2023).

Due to their rapid growth and unique physiological development, children are more sensitive to the impact of environmental exposures (World Health Organization, n.d.). Furthermore, unlike adults, children often lack autonomy, awareness, and the ability in their behaviors to make choices to safeguard their health (Ding et al., 2011; World Health Organization, n.d.). Behaviors established in childhood can also have a lasting impact on adulthood (Ortegon-Sanchez et al., 2021).

Built environments are important factors of consideration for childhood obesity prevention and control (Malacarne et al., 2022). However, the association between child health and the built environment has received less attention compared to that in adult populations (Barrett et al., 2017). This chapter surveys the literature on child obesity and relevant built environment factors and uses a pilot study to examine disparities in obesogenic built environment factors for children in East Baton Rouge Parish, Louisiana.

Literature Review of Child Obesity and the Built Environment

A number of societal and community factors related to the built environment may affect obesity rates in children: childcare and school environments, neighborhood design, availability of healthy, affordable foods and beverages, and access to safe and convenient places for physical activity (PA), which collectively affect our ability to make healthy choices (CDC, 2022). The built environment may be defined as the sum of artificially developed physical environments where people live and work (e.g., homes, buildings, streets,

DOI: 10.4324/9781032631110-11

open spaces, urban spaces, walkways, roads, parks, and infrastructure) (Gascon et al., 2016).

At the neighborhood level, Saelens et al. (2012) used Geographic Information Systems (GIS) methods to develop several neighborhood metrics for measuring obesogenic environments and found that lower child obesity risks were associated with both favorable PA environment (e.g., more conducive to walking and better access to a higher-quality park) and healthy food environment (e.g., supermarket proximity and/or lower fast-food restaurant density). Similarly, Carver et al. (2023) specified that supportive factors for PA included walkability and access to high-quality parks, and favorable nutrition environments comprised of the availability of supermarkets offering healthy food as opposed to fast-food outlets. The following offers an in-depth review of two primary ways that the obesogenic built environment can impact children's health: through PA and food environments.

Child Obesity and the Physical Activity (PA) Environment

Accessibility in a given environmental setting is one of the top themes in recent research on children's PA (Alliott et al., 2022). When examining health behaviors and outcomes for children and adolescents, it is essential to look beyond individual or family influences and consider measurements under a broader context including the neighborhood environment. Among demographic, psychosocial, behavioral, and social factors that are related to PA among adolescents, general social support was a more significant driver than parental behavior (Bauman et al., 2012). When defining the spatial scope of the built environment, relevant research has largely focused on characterizing the environment surrounding children's homes (Carver et al., 2023; Crawford et al., 2010), schools, and the facilities inside schools including those related to physical education and recreation (Van Dyke et al., 2018; Yang et al., 2023).

Moreover, Durand et al. (2012) emphasized that PA of children should include not only general physical activities but also active commuting. Active transport like walking or cycling could improve physical fitness and PA levels, though it tended to be overlooked in some related research (Smith et al., 2017). Meanwhile, active transport for children's commuting behavior may have barriers, such as the distance between school and home, or an unfriendly environment to walking and cycling (Carver et al., 2011). A recent review of the literature summarized 10 built environment factors as enablers or barriers for children's walking or other PAs, and substantiated the associations between built environment and children's weight status and health (Ortegon-Sanchez et al., 2021). The factors included residential density, street connectivity, land-use diversity, walkability, walking infrastructure, accessibility to PA facilities, accessibility to green and blue infrastructure, crime and safety, street design intervention, and social support.

Count, presence or absence, and proximity of relevant parks and recreational facilities were commonly employed to measure the environments that influence children's PA (Ding et al., 2011; Carver et al., 2023; Molina-García et al., 2022). Based on an analysis of the 100 most populous American cities, neighborhoods where most residents identified as racial or ethnic minorities had access to an average of 43% less park acreage than predominantly white neighborhoods (Foderaro, 2022). A similar disparity in park access existed in low-income neighborhoods (Foderaro and Klein, 2023). For walkability, a recent study of 500 U.S. cities reported that lower walkability correlated with increasing median income for most neighborhoods except within majority Black neighborhoods, where the relationship was reversed (Conderino et al., 2021). This suggests that a complex interplay of sociodemographic factors (such as race and socioeconomic status) likely impacts built environment disparities. Some have suggested that the concept of walking (or walkability by extension) is interpreted differently by rural residents, especially for leisure or as PA (Berry et al., 2017). Generally, rural areas had lower walkability scores and a lower prevalence of walking relative to urban areas (Berry et al., 2017; Carlson et al., 2018).

Most research, however, has relied on self-report methods for data collection and measurement. Surveys of parents and teachers have been used to measure the count or proximity of parks and recreational facilities near homes or schools (An et al., 2017; Van Dyke et al., 2018), rather than the objective measure of a number of facilities within a range or proximity to the nearest park. While specific measurements influenced the results, the most consistent associations were reported between objectively measured environmental attributes and PA, and these measures included walkability, traffic speed/volume, access/proximity to recreation facilities, land-use mix, and residential density (Ding et al., 2011). Objective measures based on tangible physical elements of the built environment are often more easily adopted and serve as a guide for practitioners when designing and planning built environments to promote desirable health outcomes (Ortegon-Sanchez et al., 2021).

Researchers studying PA began with the observation of disparities in adolescents' PA across racial and ethnic groups (Bauman et al., 2012). Colabianchi et al. (2007) found that what was considered an easy walking distance for adolescents differed by race. White adolescents, on average, considered a slightly longer time for an easy walking distance compared to African American adolescents. Therefore, compared with White adolescents, the range for PA and the accepted distance for active commuting for Blacks were more limited. The implication is that a shorter search radius needs to be employed for defining the catchment area for Black adolescents' PA. Socioeconomic position (SEP) is another widely discussed factor contributing to disparities. Low-SEP children face barriers to PA due to factors such as low quality or quantity of PA facilities in their local neighborhoods, inadequate safety and quality of public transport, and poor school facilities and activity provisions (Alliott et al., 2022).

Durand et al. (2012) utilized two sets of linear regression models to examine whether community-built environment factors promote general PA and active commuting to school by children. They found that general PA was significantly affected by walking infrastructure, lack of cul-de-sacs, and social interaction, whereas active commuting to school was significantly related to perceived crime, traffic hazards, hilliness, physical barriers, cul-de-sac connectivity, aesthetics, and walking infrastructure. However, the latter was only significant in smart growth communities. In a systematic review of the impact of the PA environment on PA and active commuting in children and adults, Smith et al. (2017) demonstrated a consistent positive influence of walkability components, quality parks and playgrounds, as well as the establishment or enhancement of active transport infrastructure on active transport engagement, PA levels, and utilization of recreational settings. However, infrastructure improvements may disproportionally benefit socioeconomically advantaged groups, resulting in a disparity in impact.

Child Obesity and the Food Environment

Another crucial dimension of the obesogenic built environment is the food environment. As highlighted in a recent study by Chambers et al. (2017), children spent more time at food outlets than structured sports and outdoor recreation locations combined. As children grow older and enter their school-age years, their access to food changes. On the one hand, they gradually become less dependent on family and have increased exposure to foods in their neighborhood environment (Shaw et al., 2023), yet their access to both healthy and less healthy foods typically remains confined to a shorter spatial range in comparison to adults (Yang et al., 2020).

Prior to assessing the food environment, an initial step is to determine how to define food providers (e.g., grocery stores, restaurants). One common approach involves using supermarkets to represent healthy food and fast-food restaurants to represent less healthy food (Egli et al., 2020; Saelens et al., 2012). The availability and accessibility of food environments may be determined by the density (including Kernel density), count, and proximity (e.g., distance from the closest facility) of specific healthy or less healthy food outlets (Egli et al., 2020; Gilliland et al., 2012; Rick et al., 2023). Similar to the methods used in assessing the PA environment, most studies on the food environment have also relied on self-reported data (e.g., surveys and interviews). However, differences in respondents' definitions of the food environment contribute to inconsistent results and make it difficult to draw meaningful comparisons on the association between children's health and the food environment (Shaw et al., 2023).

The U.S. Centers for Disease Control and Prevention (CDC) (2011) categorized food retailers into healthy and less healthy categories based on the North American Industry Classification System (NAICS). Healthy food retailers included supermarkets and larger grocery stores with more than 10 employees, fruit and vegetable markets, warehouse clubs, and supercenters (i.e., food and general merchandise). Less healthy food retailers included fast-food stores that were defined as limited-service restaurants, convenience stores, and small groceries with fewer than three employees (CDC, 2011). Based on this classification, the CDC (2011) proposed the Modified Retail Food Environment Index (mRFEI) for national census tracts, defined as a ratio of healthy food retailers to all food retailers. Another popular source for defining the food environment is the Food Access Research Atlas (FARA) from the U.S. Department of Agriculture (USDA) (2023). The FARA measures the food environment using the distance to supercenters, supermarkets, and large grocery stores.

Improving the methodology for measuring food environments is crucial to enable a more comprehensive and comparable characterization. Both measures have been widely adopted in practice. The mRFEI measurement from the CDC calculates the relative percentage, providing a more nuanced evaluation to distinguish the level of accessibility to healthy food retailers, especially for neighborhoods with no access and low access. Therefore, mRFEI is often favored over FARA for its relative measure of healthy food access (Amin et al., 2021). Some recent studies have relied on mRFEI data from the CDC collected in 2011 (Amin et al., 2021), while others have utilized an updated dataset to compute the mRFEI (McGuirt et al., 2022). Common data sources include the County Business Patterns (CBP) from the U.S. Census Bureau (Xu and Wang, 2015) and the business lists from private companies such as Reference USA and Dun & Bradstreet (Cobb et al., 2015).

Some individual-level studies have considered demographic factors such as children's race, gender, and age, when modeling the relationship between food environments, diet behavior, and childhood obesity (Egli et al., 2020). In a study in Massachusetts reported by Fiechtner et al. (2013), children living farther away from a large supermarket had a higher body mass index (BMI) than those living closer. A recent study in Minneapolis indicated that families in neighborhoods with limited access to supermarkets with fresh fruits and vegetables (termed "food deserts") experienced significantly ($p < 0.01$) lower food security at home, and their children had higher BMIs (Agarwal et al., 2022). Another study in New York City revealed that, overall, every 0.1-mile increase between a child's residence and the nearest fast-food restaurant was associated with a decrease in the probability of children being obese by approximately 0.6% (Han et al., 2020). Furthermore, the same study found that the proximity of fast-food restaurants exerted a significant negative effect on Black and Hispanic students, but not necessarily on White and Asian students, suggesting that the effect of food environment on adolescents' health may differ across racial-ethnic groups.

Findings from the literature on the association between food environment and children's health are far from conclusive. There is evidence that greater access to healthy food is positively related to children's healthier food purchase behavior and better health status, while convenient access to less healthy food negatively affects their health (Cobb et al., 2015; Zhou et al., 2021). Some studies suggest that household financial status, neighborhood deprivation level, and unhealthy dietary behaviors, rather than nutrition environment, are highly associated with children's food purchases and their health status including waist-to-height ratio (WtHR) (Egli et al., 2020). A recent review of a large number (87) of studies on the relationship between access to less healthy food like fast-food and weight-related behaviors/outcomes among children and adolescents yielded a mixed relationship (Jia et al., 2021). While a positive association between fast-food access and fast-food consumption was consistently observed, the link to other weight-related behaviors such as frequency of food consumption, dietary quality score, and PA was either mixed or nonsignificant (Jia et al., 2021). Another review of studies investigating the effects of food environment on school-aged children's obesity in North America yielded varied results with no overwhelming significant findings overall (Key et al., 2023). The inconsistent results once again may be attributable to the heterogeneity of measures and research designs as some studies used data from individuals, while others used data aggregated by areal unit.

In summary, the literature on obesogenic environment for childhood health has a large volume with mixed results. Some noticeable gaps can be identified, such as a paucity of comprehensive coverage of both PA and food environment factors, lack of objective measures of accessibility to PA facilities, and insufficient focus on geographic disparities across neighborhoods of various urbanicities and sociodemographic characteristics. These issues will be addressed in the following pilot study.

Case Study Data and Analysis

This case study focuses on East Baton Rouge Parish, the state capital of Louisiana. Parish is a county-equivalent unit in Louisiana. The study examines the built environment related to children's health with an emphasis on the disparities across areas with various urbanicity levels, minority (e.g., Black) concentration, and Area Deprivation Index (ADI) values.

The school-age child population is based on the 2021 one-year American Community Survey (ACS) (U.S. Census Bureau, 2023) and covers the 5–14-year-old children at the census block group level. Two census block groups had no children and thus were excluded from the analysis. There were 322 residential census block groups within the parish. According to the 2020 Census Urban and Rural Classification standard (https://www.census.gov/programs-surveys/geography/guidance/geo-areas/urban-rural.html), a census block group is defined as urban area (UA) if its housing density is greater than or equal to 425 housing units per square mile (HPSM), low-density fill zone (hereafter simply referred to as low density, LD) if the density is equal to 200 to 424 HPSM, and rural area (RA) if the density is less than 200 HPSM. There were 31 RA, 36 LD, and 255 UA block groups within the study area (Figure 10.1a).

Based on 2020 census data, 49.2% of the population in the study area is White and 45.2% Black. Other racial or ethnic groups are small and thus are not addressed in this study. Based on the Black population percentage (%), census block groups are categorized into five semi-open intervals (i.e., the left boundary is inclusive, while the right boundary is exclusive): 0–20, 20–40, 40–60, 60–80, and 80–100. As shown in Figure 10.1a, the Black population is concentrated in the northern part of the parish. Moreover, the parish is racially

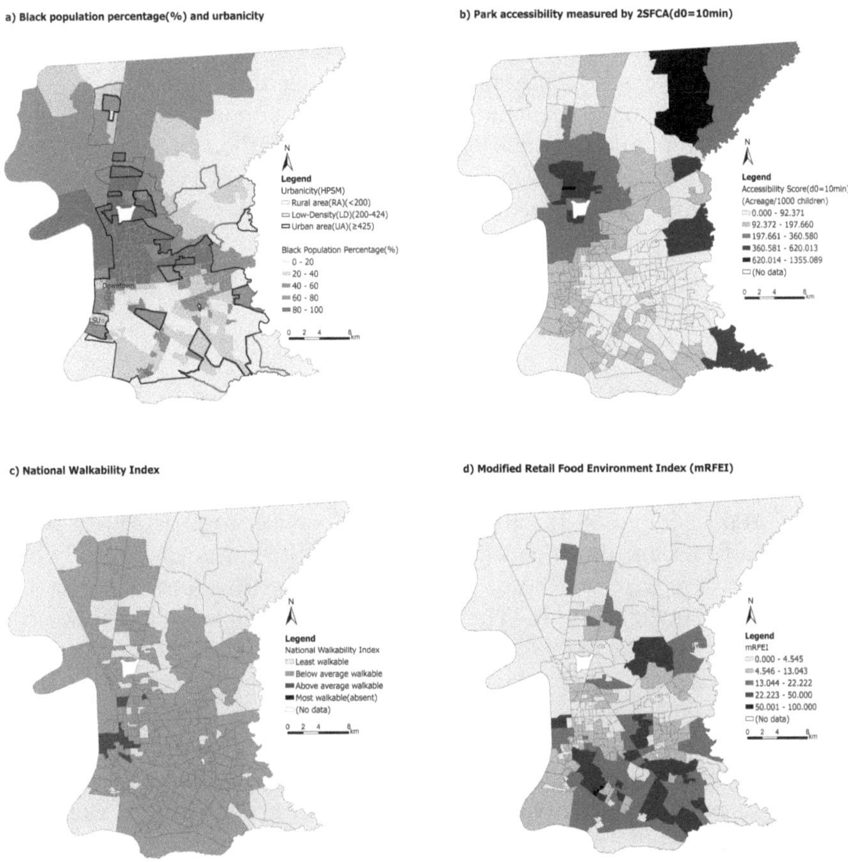

Figure 10.1 Neighborhood environmental factors by census block group in Baton Rouge, LA: (a) percent Black and urbanicity, (b) park accessibility, (c) walkability, and (d) mRFEI. Credit: Yutian Zeng.

segregated with many areas having either a low or high percent of Black population.

ADI is a comprehensive metric reflecting neighborhood socioeconomic disadvantage (Shao and Luo, 2022), developed by the University of Wisconsin School of Medicine and Public Health (2023). The 2021 ADI is ranked from 1 to 10, with 1 being the least disadvantaged and 10 the most. Like the distribution of the Black population, more disadvantaged populations are concentrated around downtown and the Louisiana State University (LSU) campus, extending toward the north and northeast. No ADI values are available for five block groups with low population.

This case study examines the built environment for children's health using three variables: accessibility of parks, walkability to capture the PA environment, and mRFEI to measure the food environment.

Based on a survey of measures of spatial accessibility of resources or services, a key element in various methods is striving to capture that users value proximity to the nearest facility and ease of securing the service, such as a ratio of supply capacity versus demand (Wang, 2021). A popular approach such as the Two-Step Floating Catchment Area (2SFCA) method captures both elements (Luo and Wang, 2003). A simple version of the 2SFCA model for accessibility at demand location (here, census block group) i is written as:

$$A_i = \sum\nolimits_{j\in(d_{ij}\leq d0)}^{n}\left[S_j / \sum\nolimits_{k\in(d_{kj}\leq d0)}^{m} D_k\right],$$

where the supply capacity of a park at location j is denoted by S_j (park acreage), the demand of census block group k (or i) is denoted by D_k (or D_i) (5–14-year-old children), and the distance (here, drive time) between them is d_{kj} (or d_{ij}). Data on park acreage was sourced from the Baton Rouge Open Data Portal operated by the BREC (2023). Road network data with speed archives in Louisiana downloaded from the Open Street Map data via Python OSMnx package were used to calculate the shortest drive time (d_{kj} or d_{ij}) in ArcGIS Pro (Boeing, 2017). The calibration was implemented using the ArcGIS toolkit for the 2SFCA method (Wang and Liu, 2024). Since the accessibility values are small, we inflate them 1000 times so that they are interpreted as acres per 1,000 children. A higher score suggests better access to these recreational spaces. We conducted a sensitivity analysis by experimenting with various catchment areas (d0) such as 5, 10, 15, and 20 minutes. The spatial patterns of park accessibility are largely consistent across the four scenarios, so we selected a middle-point value, that is, $d0 = 10$ min, as a representative case. As shown in Figure 10.1b, north Baton Rouge generally outscores the south in park accessibility.

Walkability is represented by the National Walkability Index (NWI) from the EPA (2021). The NWI is a weighted average of intersection density, proximity to transit stops, employment, and employment and household mix, defined in the Smart Location Database (SLD, available via https://www.epa.gov/smartgrowth/smart-location-mapping). The block groups are assigned their NWI scores on a scale of 1–20, categorized as least walkable (1–5.75), below-average walkable (5.76–10.5), above-average walkable (10.51–15.25), and most walkable (15.26–20). The NWI data for the year 2021 was aggregated to 2010 census block groups, so we converted it to 2020 census block groups using "areal weighting interpolator" (Wang & Liu, 2024, p. 77). The mean NWI value in Baton Rouge was 7.182, which is considered "below average" walkable. As shown in Figure 10.1c, most areas in Baton Rouge are in the least walkable or below-average walkable categories. Only a limited number (14 of 322) of block groups around downtown and scattered around parks have above-average walkability. None reaches the standard of most walkable. This reflects an overall poor walkability throughout the parish.

Based on real-time updated business lists from the Data Axle USA API (https://www.data-axle.com/our-data/apis/), the mRFEI values are calculated using the 2020 NAICS codes. For example, 445230 and 455211 represent healthy food retailers, and 722513 and 445131 represent less healthy food retailers. However, 445110 covers both healthy (e.g., supermarkets, large grocery stores) and less healthy food retailers (e.g., gas station stores, small grocery stores), which are further differentiated by a store's employment size sourced from the Buzzfile database (https://www.buzzfile.com/Home/Basic). Fifty-three data points lacking employment size information are excluded from the analysis. The range of mRFEI is from 0 to 100, and a higher mRFEI indicates a greater presence of healthy food retail stores. Based on the 2011 CDC national census tract-level data, the average mRFEI is 12.765 (https://www.cdc.gov/nutrition/resources-publications/). The mean mRFEI value for East Baton Rouge Parish is 10.063 and suggests a generally low level of healthy food accessibility. Figure 10.1d shows that areas in the southern part of the Parish exhibit higher mRFEI values and thus a healthier food environment. Regions with lower ADI tend to have a healthier food environment. Such an observation needs to be confirmed by more rigorous analysis in the following section.

Disparities in the Built Environment by Urbanicity, Black Concentration, and ADI Levels

The statistical difference in each built environment variable (i.e., park accessibility, walkability, and food environment) across multiple categories of urbanicity, Black population concentration, and ADI is accomplished using a regression model with dummy variables. The regression results are summarized in Table 10.1. For urbanicity, its association is statistically significant on each of the three factors. The average park accessibility score is the highest in rural areas (RA) (273.4), declines by 72.9 to 200.5 in LD areas, and by 145.4 to 128.1 in UA. The average walkability index is the lowest in RA (4.8), increases by 1.2 to 6.0 in LD, and by 2.8 to 7.6 in UA. The average mRFEI is 2.0 in RA, increases by 7.8 to 9.8 in LD, and by 9.0 to 11.0 in UA, respectively. Overall, urbanized areas enjoy more favorable environments in healthy food accessibility and walkability but not park access. Note that this study uses park acreage to measure park size, which consequently influences accessibility, and favors rural residents as they are generally closer to large parks in the study area.

There is no significant variation in walkability across different concentration levels of Black populations. However, areas with the highest concentration (80–100%) of Black population exhibited significantly higher park accessibility and less healthy food environments. One likely reason for the high park accessibility in these areas is the colocation of the high concentration of Black population and relatively high park access in north rural Baton Rouge. The benefits of park access in these areas may not be realized if children rely on their parents to provide transportation. Additionally, these rural areas suffer from a less healthy food environment.

Table 10.1 Disparity in park accessibility, walkability, and food environment in Baton Rouge.

	Park access (d_0 =10 min) (*acreage/1000 children*)			*Walkability index*			*Food environment* (*mRFEI*)		
	Estimate	*Std. error*	*P value*	*Estimate*	*Std. error*	*P value*	*Estimate*	*Std. error*	*P value*
Urbanicity									
Reference:									
Rural area (RA)	273.433	27.337	0.000***	4.827	0.325	0.000***	2.007	2.065	0.332
Low-density zones (LD)	−72.893	36.733	0.048*	1.188	0.436	0.007**	7.756	2.774	0.005**
Urban area (UA)	−145.414	28.867	0.000***	2.793	0.343	0.000***	9.027	2.180	0.000***
Percent Black									
Reference:0–20	127.184	15.690	0.000***	7.198	0.206	0.000***	14.479	1.110	0.000***
20–40	−3.500	25.863	0.892	0.134	0.340	0.694	−1.548	1.830	0.398
40–60	−5.620	28.909	0.846	−0.299	0.380	0.432	−3.723	2.046	0.070
60–80	−2.466	27.698	0.929	0.344	0.364	0.346	−4.309	1.960	0.029*
80–100	82.581	22.129	0.000***	−0.170	0.291	0.559	−10.773	1.566	0.000***
ADI									
Reference: 1	130.004	18.472	0.000***	7.469	0.247	0.000***	16.239	1.323	0.000***
2	−45.269	28.434	0.112	−0.279	0.380	0.464	−3.129	2.036	0.125
3	40.641	30.909	0.190	−0.417	0.414	0.315	−3.546	2.213	0.110
4	−59.547	35.718	0.097	0.111	0.478	0.817	−3.071	2.558	0.231
5	18.539	32.165	0.565	−0.325	0.430	0.451	−9.154	2.303	0.000***
6	−43.598	43.320	0.315	−0.719	0.580	0.216	−6.411	3.102	0.040*
7	59.851	46.179	0.196	−0.437	0.618	0.480	−8.995	3.307	0.007**
8	46.937	32.900	0.155	−0.943	0.440	0.033*	−11.710	2.356	0.000***
9	110.297	35.169	0.002**	−0.768	0.471	0.104	−11.859	2.518	0.000***
10	88.430	29.641	0.003**	0.126	0.397	0.751	−12.838	2.123	0.000***

*Significant at 0.05, ** significant at 0.01, *** significant at 0.001.

Similar disparity patterns are observed across the ADI values: the variation of walkability is negligible, there is better park access for areas with ADI ≥ 9, and the mRFEI value consistently worsens after passing its middle range (i.e., ADI ≥ 5). In other words, access to healthy food significantly decreases in areas with concentrated disadvantages, particularly in areas with the highest disadvantage levels when ADI is 8–10. Once again, the relative advantage in highly disadvantaged areas for park access via drive time needs to be interpreted with caution.

Concluding Remarks

Overall, the obesogenic built environment factors explored in this case study are not conducive to children's health in Baton Rouge, LA with scores below the national average for walkability and healthy food availability. The spatial patterns of these three factors lacked consistency.

The highest park accessibility scores were observed in rural areas and locations with the highest Black concentration (80–100%) or the highest ADI values (9–10). This is counterintuitive and in disagreement with most of the literature (Foderaro, 2022; Foderaro and Klein, 2023). The current measure of park accessibility uses the park's acreage to assess capacity and gives an edge to rural residents that are next to large parks in the study area. Areas with higher Black percentage or ADI values are generally closer to parks and thus enjoy better physical accessibility. However, a larger area does not necessarily equate to more recreational equipment or events available to children. Moreover, the accessibility of parks is based on a catchment area defined by a fixed drive time range. Residents' spatial proximity to such facilities does not necessarily convert to actual access given the complexity of transportation modes by children. A refined measure of park access is needed in a follow-up in-depth study.

For walkability, there is evidence of an urban advantage, as indicated by the literature (Carlson et al., 2018), but no noticeable variability across various Black population % or ADI values. The latter may be attributable to the complex overlapping effects of sociodemographic factors on walkability as revealed in Conderino et al. (2021). Moreover, the mere presence of walkable infrastructure does not guarantee its usage. As suggested by the literature review, additional factors such as traffic, crime, street view, etc. can all play important roles in walkability. In terms of food environment (mRFEI), the least favorable conditions are observed in rural areas, neighborhoods with higher concentrations (>60%) of Blacks, and areas with higher ADI values (≥5). This result is consistent with most of the literature (Glover et al., 2022; Jiang et al., 2023).

In summary, the obesity prevalence among children in the United States is largely attributable to disparities in obesogenic environment across geographic areas and sociodemographic groups. Assessment of the environment needs to include two primary pathways, namely PA and food intake. The list of environmental factors conducive to PA is long in the literature, and this pilot study selects accessibility to parks and walkability, two of the most consistent variables associated with PA. For food environment in the United States, most studies either use the CDC's mRFEI or the USDA's FARA. This study chooses mRFEI over FARA for its relative measure of healthy food access. All three variables are considered objective measures, and thus the results may shed light on actionable public policy. Reducing health disparities is a major vehicle for achieving the improvement of overall health outcomes in the United States (Brown et al., 2019). For evaluating disparities across geographic areas, our analysis emphasizes urbanity, a common theme in public health studies (Wang, 2020), and for inequality across sociodemographic factors, we use Black concentration levels to capture the racial-ethnic domain and ADI as a comprehensive index for socioeconomic structure.

Funding

This study was supported in part by the Public University Partnership Program at the Louisiana Department of Health, Bureau of Health Services Financing (Senlin Chen as PI).

Disclaimer

The content presented in this book chapter is the sole responsibility of the authors and does not necessarily represent the official views of the Louisiana Department of Health.

References

Agarwal, Sarthak, Angela R Fertig, Amanda C Trofholz, Allan D Tate, Jenna Robinson, and Jerica M Berge. 2022. "Exploring the Associations between Neighbourhood Food Environment, Household Food Insecurity and Child Weight-Related Outcomes in Socio-Economically and Racially/Ethnically Diverse Families." *Public Health Nutrition* 25 (12): 3538–47. https://doi.org/10.1017/S1368980022002130

Alliott, Olivia, Mairead Ryan, Hannah Fairbrother, and Esther Van Sluijs. 2022. "Do Adolescents' Experiences of the Barriers to and Facilitators of Physical Activity Differ by Socioeconomic Position? A Systematic Review of Qualitative Evidence." *Obesity Reviews* 23 (3): e13374. https://doi.org/10.1111/obr.13374

Amin, Modhurima Dey, Syed Badruddoza, and Jill J. McCluskey. 2021. "Predicting Access to Healthful Food Retailers with Machine Learning." *Food Policy* 99 (February): 101985. https://doi.org/10.1016/j.foodpol.2020.101985

An, Ruopeng, Yan Yang, and Kaigang Li. 2017. "Residential Neighborhood Amenities and Physical Activity Among U.S. Children with Special Health Care Needs." *Maternal and Child Health Journal* 21 (5): 1026–36. https://doi.org/10.1007/s10995-016-2198-3

Barrett, Millie, Sarah Crozier, Daniel Lewis, Keith Godfrey, Sian Robinson, Cyrus Cooper, Hazel Inskip, Janis Baird, and Christina Vogel. 2017. "Greater Access to Healthy Food Outlets in the Home and School Environment Is Associated with Better Dietary Quality in Young Children." *Public Health Nutrition* 20 (18): 3316–25. https://doi.org/10.1017/S1368980017002075

Bauman, Adrian E, Rodrigo S Reis, James F Sallis, Jonathan C Wells, Ruth Jf Loos, and Brian W Martin. 2012. "Correlates of Physical Activity: Why Are Some People

Physically Active and Others Not?" *The Lancet* 380 (9838): 258–71. https://doi.org/10.1016/S0140-6736(12)60735-1

Berry, Narelle M., Neil T. Coffee, Rebecca Nolan, James Dollman, and Takemi Sugiyama. 2017. Neighbourhood Environmental Attributes Associated with Walking in South Australian Adults: Differences between Urban and Rural Areas. *International Journal of Environmental Research and Public Health* 14: 965.

Boeing, G. 2017. "New Methods for Acquiring, Constructing, Analyzing, and Visualizing Complex Street Networks." *Computers, Environment and Urban Systems*, no. 65: 126–39. https://doi.org/10.1016/j.compenvurbsys.2017.05.004

BREC. 2023. "BREC Park Amenities." Open Data BR. 2023. https://data.brla.gov/Culture-and-Recreation/BREC-Park-Amenities/phg8-g77c/about_data

Brown, Arleen F., Grace X. Ma, Jeanne Miranda, Eugenia Eng, Dorothy Castille, Teresa Brockie, Patricia Jones, Collins O. Airhihenbuwa, Tilda Farhat, Lin Zhu, and Chau Trinh-Shevrin. 2019. Structural Interventions to Reduce and Eliminate Health Disparities. *American Journal of Public Health* 109(S1):S72–S78. https://doi.org/10.2105/AJPH.2018.304844

Carlson, Susan A., Geoffrey P. Whitfield, Erin L. Peterson, Emily N. Ussery, Kathleen B. Watson, David Berrigan, Janet E. Fulton. 2018. Geographic and Urban–Rural Differences in Walking for Leisure and Transportation. *American Journal of Preventive Medicine* 55(6): 887–95.

Carver, Alison, Anna F Timperio, Kylie D Hesketh, Nicola D Ridgers, Jo L Salmon, and David A Crawford. 2011. "How Is Active Transport Associated with Children's and Adolescents' Physical Activity over Time?" *International Journal of Behavioral Nutrition and Physical Activity* 8 (1): 126. https://doi.org/10.1186/1479-5868-8-126

Carver, Alison, Ester Cerin, Muhammad Akram, James F. Sallis, Kelli L. Cain, Lawrence D. Frank, Carrie M. Geremia, Terry L. Conway, Karen Glanz, and Brian E. Saelens. 2023. "Associations of Home and Neighborhood Environments with Children's Physical Activity in the U.S.-Based Neighborhood Impact on Kids (NIK) Longitudinal Cohort Study." *International Journal of Behavioral Nutrition and Physical Activity* 20 (1): 9. https://doi.org/10.1186/s12966-023-01415-3

CDC. 2011. "*Census Tract Level State Maps of the Modified Retail Food Environment Index (mRFEI)*." Center for Disease Control and Prevention. https://www.cdc.gov/obesity/downloads/census-tract-level-state-maps-mrfei_TAG508.pdf

CDC. 2022. "Childhood Overweight & Obesity." April 1, 2022. https://www.cdc.gov/obesity/childhood/index.html

Chambers, T., A. L. Pearson, I. Kawachi, Z. Rzotkiewicz, J. Stanley, M. Smith, M. Barr, C. Ni Mhurchu, and L. Signal. 2017. "Kids in Space: Measuring Children's Residential Neighborhoods and Other Destinations Using Activity Space GPS and Wearable Camera Data." *Social Science & Medicine* 193 (November): 41–50. https://doi.org/10.1016/j.socscimed.2017.09.046

Cobb, Laura K., Lawrence J. Appel, Manuel Franco, Jessica C. Jones-Smith, Alana Nur, and Cheryl A. M. Anderson. 2015. "The Relationship of the Local Food Environment with Obesity: A Systematic Review of Methods, Study Quality, and Results: The Local Food Environment and Obesity." *Obesity* 23 (7): 1331–44. https://doi.org/10.1002/oby.21118

Colabianchi, Natalie, Marsha Dowda, Karin A Pfeiffer, Dwayne E Porter, Maria Almeida, and Russell R Pate. 2007. "Towards an Understanding of Salient Neighborhood Boundaries: Adolescent Reports of an Easy Walking Distance and Convenient Driving Distance." *International Journal of Behavioral Nutrition and Physical Activity* 4 (1): 66. https://doi.org/10.1186/1479-5868-4-66

Conderino, Sarah E., Justin M. Feldman, Benjamin Spoer, Marc N. Gourevitch, Lorna E. Thorpe. 2021. Social and Economic Differences in Neighborhood Walkability Across 500 U.S. Cities. *American Journal of Preventive Medicine* 61(3):394–401.

Crawford, D, V Cleland, A Timperio, J Salmon, N Andrianopoulos, R Roberts, B Giles-Corti, L Baur, and K Ball. 2010. "The Longitudinal Influence of Home and Neighbourhood Environments on Children's Body Mass Index and Physical Activity over 5 Years: The CLAN Study." *International Journal of Obesity* 34 (7): 1177–87. https://doi.org/10.1038/ijo.2010.57

Ding, Ding, James F. Sallis, Jacqueline Kerr, Suzanna Lee, and Dori E. Rosenberg. 2011. "Neighborhood Environment and Physical Activity Among Youth." *American Journal of Preventive Medicine* 41 (4): 442–55. https://doi.org/10.1016/j.amepre.2011.06.036

Durand, Casey P., Genevieve F. Dunton, Donna Spruijt-Metz, and Mary Ann Pentz. 2012. "Does Community Type Moderate the Relationship between Parent Perceptions of the Neighborhood and Physical Activity in Children?" *American Journal of Health Promotion* 26 (6): 371–80. https://doi.org/10.4278/ajhp.100827-QUAN-290

Van Dyke, Miriam E., Patricia C. Cheung, Padra Franks, and Julie A. Gazmararian. 2018. "Socioeconomic and Racial/Ethnic Disparities in Physical Activity Environments in Georgia Elementary Schools." *American Journal of Health Promotion* 32 (2): 453–63. https://doi.org/10.1177/0890117117717016

Egli, Victoria, Matthew Hobbs, Jordan Carlson, Niamh Donnellan, Lisa Mackay, Daniel Exeter, Karen Villanueva, Caryn Zinn, and Melody Smith. 2020. "Deprivation Matters: Understanding Associations between Neighbourhood Deprivation, Unhealthy Food Outlets, Unhealthy Dietary Behaviours and Child Body Size Using Structural Equation Modelling." *Journal of Epidemiology and Community Health* 74 (5): 460–66. https://doi.org/10.1136/jech-2019-213159

Fiechtner, Lauren, Jason Block, Dustin T. Duncan, Matthew W. Gillman, Steven L. Gortmaker, Steven J. Melly, Sheryl L. Rifas-Shiman, and Elsie M. Taveras. 2013. "Proximity to Supermarkets Associated with Higher Body Mass Index among Overweight and Obese Preschool-Age Children." *Preventive Medicine* 56 (3–4): 218–21. https://doi.org/10.1016/j.ypmed.2012.11.023

Foderaro, Lisa W. 2022. *The Power of Parks to Address Climate Change: A Special Report*. available via https://www.tpl.org/parks-address-climate-change-report

Foderaro, Lisa W. and Will Klein. 2023. *The Power of Parks to Promote Health: A Special Report*. available via https://www.tpl.org/parks-promote-health-report

Gascon, Mireia, Martine Vrijheid, and Mark J. Nieuwenhuijsen. 2016. "The Built Environment and Child Health: An Overview of Current Evidence." *Current Environmental Health Reports* 3 (3): 250–57. https://doi.org/10.1007/s40572-016-0094-z

Gilliland, Jason A., Claudia Y. Rangel, Martin A. Healy, Patricia Tucker, Janet E. Loebach, Paul M. Hess, Meizi He, Jennifer D. Irwin, and Piotr Wilk. 2012. "Linking Childhood Obesity to the Built Environment: A Multi-Level Analysis of Home and School Neighbourhood Factors Associated With Body Mass Index." *Canadian Journal of Public Health* 103: S15–21.

Glover, Bailey, Liang Mao, Yujie Hu, and Jiawen Zhang. 2022. Enhancing the Retail Food Environment Index (RFEI) with Neighborhood Commuting Patterns: A Hybrid Human-Environment Measure. International Journal of Environmental Research and Public Health 19(17):10798. https://doi.org/10.3390/ijerph191710798

Han, Jeehee, Amy Ellen Schwartz, and Brian Elbel. 2020. "Does Proximity to Fast Food Cause Childhood Obesity? Evidence from Public Housing." *Regional Science and Urban Economics*, no. 84. https://doi.org/10.1016/j.regsciurbeco.2020.103565

Jia, Peng, Miyang Luo, Yamei Li, Ju-Sheng Zheng, Qian Xiao, and Jiayou Luo. 2021. "Fast-food Restaurant, Unhealthy Eating, and Childhood Obesity: A Systematic Review and Meta-analysis." *Obesity Reviews* 22. https://doi.org/10.1111/obr.12944

Jiang, Qianxia, Debarchana Ghosh, Sandro Steinbach, and Kristen Cooksey Stowers. 2023. A Longitudinal Assessment of Racial and Ethnic Inequities in Food Environment Exposure and Retail Market Concentration. *Public Health Nutrition* 26(9):1850–61. https://doi.org/10.1017/S1368980023001179

Key, Johanna, Donna Burnett, Jeganathan Ramesh Babu, and Thangiah Geetha. 2023. "The Effects of Food Environment on Obesity in Children: A Systematic Review." *Children* 10 (1): 98. https://doi.org/10.3390/children10010098

Lartey, Stella T., Wasantha P. Jayawardene, Stephanie L. Dickinson, Xiwei Chen, Nana Gletsu-Miller, and David K. Lohrmann. 2023. "Evaluation of Unintended Consequences of COVID-19 Pandemic Restrictions and Obesity Prevalence Among Youths." *JAMA Network Open* 6 (7): e2323596. https://doi.org/10.1001/jamanetworkopen.2023.23596

Luo, Wei, and Fahui Wang. 2003. "Measures of Spatial Accessibility to Health Care in a GIS Environment: Synthesis and a Case Study in the Chicago Region." *Environment and Planning. B, Planning & Design* 30 (6): 865–84. https://doi.org/10.1068/b29120

Malacarne, Diego, Evangelos Handakas, Oliver Robinson, Elisa Pineda, Marc Saez, Leda Chatzi, and Daniela Fecht. 2022. "The Built Environment as Determinant of Childhood Obesity: A Systematic Literature Review." *Obesity Reviews* 23 (S1): e13385. https://doi.org/10.1111/obr.13385

McGuirt, Jared T., Stephanie B. Jilcott Pitts, Jeffrey D. Labban, Elizabeth T. Anderson Steeves, Lindsey Haynes-Maslow, Savanna Henry, and Alison Gustafson. 2022. "Evidence of Geospatial and Socioeconomic Disparities in Access to Online Grocery Shopping for Fresh and Frozen Produce in North Carolina." *Journal of the Academy of Nutrition and Dietetics* 122 (11): 2106–14. https://doi.org/10.1016/j.jand.2022.05.008

Molina-García, Javier, Cristina Menescardi, Isaac Estevan, and Ana Queralt. 2022. "Associations between Park and Playground Availability and Proximity and Children's Physical Activity and Body Mass Index: The BEACH Study." *International Journal of Environmental Research and Public Health* 19 (1): 250. https://doi.org/10.3390/ijerph19010250

Ogden, Cynthia L., Margaret D. Carroll, Tala H. Fakhouri, Craig M. Hales, Cheryl D. Fryar, Xianfen Li, and David S. Freedman. 2018. "Prevalence of Obesity Among Youths by Household Income and Education Level of Head of Household – United States 2011–2014." *MMWR. Morbidity and Mortality Weekly Report* 67 (6): 186–89. https://doi.org/10.15585/mmwr.mm6706a3

Ortegon-Sanchez, Adriana, Rosemary R. C. McEachan, Alexandra Albert, Chris Cartwright, Nicola Christie, Ashley Dhanani, Shahid Islam, Marcella Ucci, and Laura Vaughan. 2021. "Measuring the Built Environment in Studies of Child Health—A Meta-Narrative Review of Associations." *International Journal of Environmental Research and Public Health* 18 (20): 10741. https://doi.org/10.3390/ijerph182010741

Rick, Christopher, Jeehee Han, Brian Elbel, and Amy Ellen Schwartz. 2023. "The Link between Gentrification, Children's Egocentric Food Environment, and Obesity." *Housing Policy Debate* 33 (1): 85–106. https://doi.org/10.1080/10511482.2022.2125788

Saelens, Brian E., James F. Sallis, Lawrence D. Frank, Sarah C. Couch, Chuan Zhou, Trina Colburn, Kelli L. Cain, James Chapman, and Karen Glanz. 2012. "Obesogenic Neighborhood Environments, Child and Parent Obesity: The Neighborhood Impact on Kids Study." *American Journal of Preventive Medicine* 42 (5): e57–64. https://doi.org/10.1016/j.amepre.2012.02.008

Shao, Yaxiong, and Wei Luo. 2022. "Supply-Demand Adjusted Two-Steps Floating Catchment Area (SDA-2SFCA) Model for Measuring Spatial Access to Health Care." *Social Science & Medicine* 296 (March): 114727. https://doi.org/10.1016/j.socscimed.2022.114727

Shaw, Sarah, Millie Barrett, Calum Shand, Cyrus Cooper, Sarah Crozier, Dianna Smith, Mary Barker, and Christina Vogel. 2023. "Influences of the Community and Consumer Nutrition Environment on the Food Purchases and Dietary Behaviors of Adolescents: A Systematic Review." *Obesity Reviews* 24 (7): e13569. https://doi.org/10.1111/obr.13569

Smith, Melody, Jamie Hosking, Alistair Woodward, Karen Witten, Alexandra MacMillan, Adrian Field, Peter Baas, and Hamish Mackie. 2017. "Systematic Literature Review of Built Environment Effects on Physical Activity and Active Transport – an Update and New Findings on Health Equity." *International Journal of Behavioral Nutrition and Physical Activity* 14 (1): 158. https://doi.org/10.1186/s12966-017-0613-9

Stierman, Bryan, Joseph Afful, Margaret D. Carroll, Te-Ching Chen, Orlando Davy, and Steven Fink. 2021. "National Health and Nutrition Examination Survey 2017–March 2020 Prepandemic Data Files—Development of Files and Prevalence Estimates for Selected Health Outcomes." *National Health Statistics Reports*, no. 158. 1-21. https://www.cdc.gov/nchs/data/nhsr/nhsr158-508.pdf

U.S. Census Bureau. 2023. "2021 ACS 1-Year Estimates." 2023. https://www.census.gov/programs-surveys/acs/technical-documentation/table-and-geography-changes/2021/1-year.html

U.S. Environmental Protection Agency. 2021. "*National Walkability Index User Guide and Methodology*." U.S. Environmental Protection Agency. June 16, 2021. (https://www.epa.gov/smartgrowth/national-walkability-index-user-guide-and-methodology

University of Wisconsin School of Medicine and Public Health. 2023. "Area Deprivation Index." 2023. https://www.neighborhoodatlas.medicine.wisc.edu/

USDA. 2023. "Food Access Research Atlas." USDA Economic Research Service. September 27, 2023. https://www.ers.usda.gov/data-products/food-access-research-atlas/

Wang, Fahui. 2020. Why Public Health Needs GIS? A Methodological Review. *Annals of GIS* 26: 1–12. https://doi.org/10.1080/19475683.2019.1702099

Wang, Fahui, and Lingbo Liu. 2024. *Computational Methods and GIS Applications in Social Science*. 3rd ed. Boca Raton, FL: CRC Press.

Wang, Fahui. 2021. "From 2SFCA to i2SFCA: Integration, Derivation and Validation." *International Journal of Geographical Information Science* 35 (3): 628–38. https://doi.org/10.1080/13658816.2020.1811868

World Health Organization. n.d. "Children's Environmental Health." Accessed October 1, 2023. https://www.who.int/health-topics/children-environmental-health#tab=tab_2

Xu, Yanqing, and Fahui Wang. 2015. "Built Environment and Obesity by Urbanicity in the U.S." *Health & Place* 34 (July): 19–29. https://doi.org/10.1016/j.healthplace.2015.03.010

Yang, Jue, Janani Rajbhandari-Thapa, Lan Mu, Donglan Zhang, and Kiran Thapa. 2023. "Physical Activity Environment Inside and Outside School and Youth Fitness: A Spatial Analysis at the Level of Elementary School Attendance Zone in Georgia." *Indoor and Built Environment* 32 (4): 719–33. https://doi.org/10.1177/1420326X221135337

Yang, Meng, Haoluan Wang, and Feng Qiu. 2020. "Neighbourhood Food Environments Revisited: When Food Deserts Meet Food Swamps." *The Canadian Geographer / Le Géographe Canadien* 64 (1): 135–54. https://doi.org/10.1111/cag.12570

Zhou, Qian, Li Zhao, Longhao Zhang, Qian Xiao, Tong Wu, Tommy Visscher, Jinfeng Zhao, et al. 2021. "Neighborhood Supermarket Access and Childhood Obesity: A Systematic Review." *Obesity Reviews* 22 (S1). https://doi.org/10.1111/obr.12937

11 Health and Equity in Age-Friendly Cities and Communities

A Global Geographic Perspective

Sarah Nelson

Introduction

Globally, the movement for Age-Friendly Cities and Communities (AFCC) has gained momentum since the World Health Organization championed the AFCC framework and initiated the Global Network for AFCC in 2010 (World Health Organization 2024). Many cities and communities around the world have embraced the concept, and research has proliferated. This is driven by an increasingly aged global population: according to the World Health Organization (2023), "the proportion of older people in the world is projected to reach nearly 12% in 2030 and 16% in 2050" (2). Although high-income countries have higher proportions of older people relative to the population, due to larger population numbers in general, the number of older people in low- and middle-income countries is highest and projected to grow (World Health Organization 2023). Much of the world's population growth, including for older people, is projected to occur in urban areas. The UN Habitat World Cities Report notes that "the future of humanity is undoubtedly urban," and calls for "cities for all" in which people at all life stages are included in urban planning (2022, xv). The AFCC movement needs to also be responsive to the rapid growth of urbanization around the world. Research on AFCC in the past decade has focused on the intersection of growing urbanization and aging populations. Scholars point to inequities between cities, for example between cities in high-income countries versus low-income countries, as well as inequities within cities that impact older people through the processes of gentrification, economic austerity, and the capitalist bent toward economic growth and productivity (Steels 2015; Buffel, Phillipson & Scharf, 2012a; Buffel and Phillipson 2016). These inequities affect different groups of older people in different ways and require particular attention when developing and implementing age-friendly policies if these policies are to make lasting and consequential changes in the lives of older people around the world (Herman et al., 2020; Nelson & Rosenberg, 2021).

Specific areas of focus of AFCC plans generally include adjustments to "the physical environment, housing, transportation, economic determinants, social determinants, health and social services, communication, information, social

DOI: 10.4324/9781032631110-12

connectedness, social participation, physical activity, civic engagement, personal determinants and further education" (Steels 2015, 47). In many cases, the core elements that receive the greatest attention in age-friendly planning and research have to do with physical accessibility: for example, housing, mobility, and outdoor spaces and buildings, with the participation of older people in civic and community life also highlighted as important (Fitzgerald and Caro 2014). Success in improving the age-friendliness of these various components of cities and people's lives requires attention to the interconnections among the elements of the physical and social environment (e.g., linking transportation initiatives with social connectedness or accessible housing with health care services), intersectoral collaboration between stakeholders and various partners, and political commitment, including the meaningful involvement of older persons in decision making (Torku, Chan, and Yung 2020; Steels 2015). Improvement in each of these areas on their own is less likely to make a significant difference in the lives of older people in a city nor is a siloed approach in which various departments act independently of one another. In addition, decisions made on behalf of older people without their active involvement are unlikely to make meaningful change.

In response to years of advocacy by the WHO and other stakeholders, the United Nations declared 2021–2030 to be the UN Decade of Healthy Aging. The stated goals for this decade are to:

- Change how we think, feel, and act toward age and aging.
- Ensure communities that foster the abilities of older people.
- Deliver person-centered, integrated care, and primary health services that are responsive to older people.
- Provide access to long-term care for older people who need it (World Health Organization 2023).

These goals suggest a shift in age-friendly thinking, away from an overly heavy emphasis on physical accessibility and affordability toward more strengths-based thinking that avoids the pitfalls of ageism. Changing how we think, feel, and act toward older people means reflecting on assumptions that may have been inherent in previous policies. Focusing on the abilities and unique contributions of older people as well as on person-centered care avoids dehumanizing or paternalistic approaches to the functions of older people in societies around the world. Ensuring access to long-term care also avoids an overemphasis on "active aging," which has been criticized for promoting an at-times unrealistic ideal of physical activity in old age, and disadvantaging the frail, ill, or those living with disabilities (Paz, Doron, and Tur-Sinai 2018; Boudiny 2013).

Thus, it appears that AFCC approaches, at least at the level of international organizations such as the World Health Organization and the United Nations, have taken a turn toward equity, person-centeredness, and inclusion and have taken lessons from previous scholarship, planning, and policy to heart. As the global movement for age-friendliness gains momentum, however, there are some areas which merit ongoing critical attention, building on the work of previous scholars and bringing research from diverse geographies and disciplines into conversation to identify the best paths forward. This chapter engages with critical research on AFCC to delve into areas of ongoing debate and offer suggestions for future research and policy worldwide.

Geographic Specificity of Age-Friendly Policy

The World Health Organization's Global Network for AFCC was launched in 2010, with the mission "to stimulate and enable cities and communities around the world to become increasingly age-friendly" (World Health Organization 2024). Any city or community from around the world can apply for membership in this network, which facilitates connections and collaborations among different cities and communities and promotes the sharing of programs and best practices. Membership in the network is free of cost; however, an aspiring member must demonstrate that they are committed to developing their capacity and potential to become an age-friendly city or community.

This global network now includes 1542 cities and communities in 51 different countries (World Health Organization 2024) (Figure 11.1). However, membership shows particular geographic patterns. Apart from South Korea (which has a number of cities that are part of the WHO network), Asian countries contained only 7% of network members as of 2018 (Woo and Choi 2020). India – one of the most rapidly growing populations of the world, and one of the most quickly urbanizing – also has no members. Australia has few members, while New Zealand has many. Chile has by far the highest concentration of members in all of South America. European members are concentrated in the West, and in North America, although the continent has the highest number of members overall, AFCCs are concentrated along the eastern part of the continent. There are currently no members across the continent of Africa, in spite of a growth of overall network members from 799 in 2018 to 1542 in 2023 (World Health Organization 2024), and the COVID-19 pandemic's disproportionate impact on older individuals across Africa, like everywhere else in the world (WHO Regional Office for Africa 2021).

Underrepresentation of specific regions in the Global Network for AFCC, in particular low- and middle-income countries and those with cultures that differ from those dominant in Western European and North American societies, has been pointed out by scholars (Steels 2015; Woo and Choi 2020). The AFCC framework, in spite of efforts to incorporate more flexibility and adaptability in recent years, has been criticized for having an overly Western orientation (Torku, Chan, and Yung 2020; Xiang et al. 2020; Buffel and Phillipson 2016). This remains reflected in the uneven geographic distribution of membership in the WHO global network. Although much of the AFCC literature has tended to emphasize aging as a problem of the developed world (Fitzgerald and Caro 2014), the need to accommodate aging populations extends well beyond high-income countries. According to Steels, "in five decades, over 80%

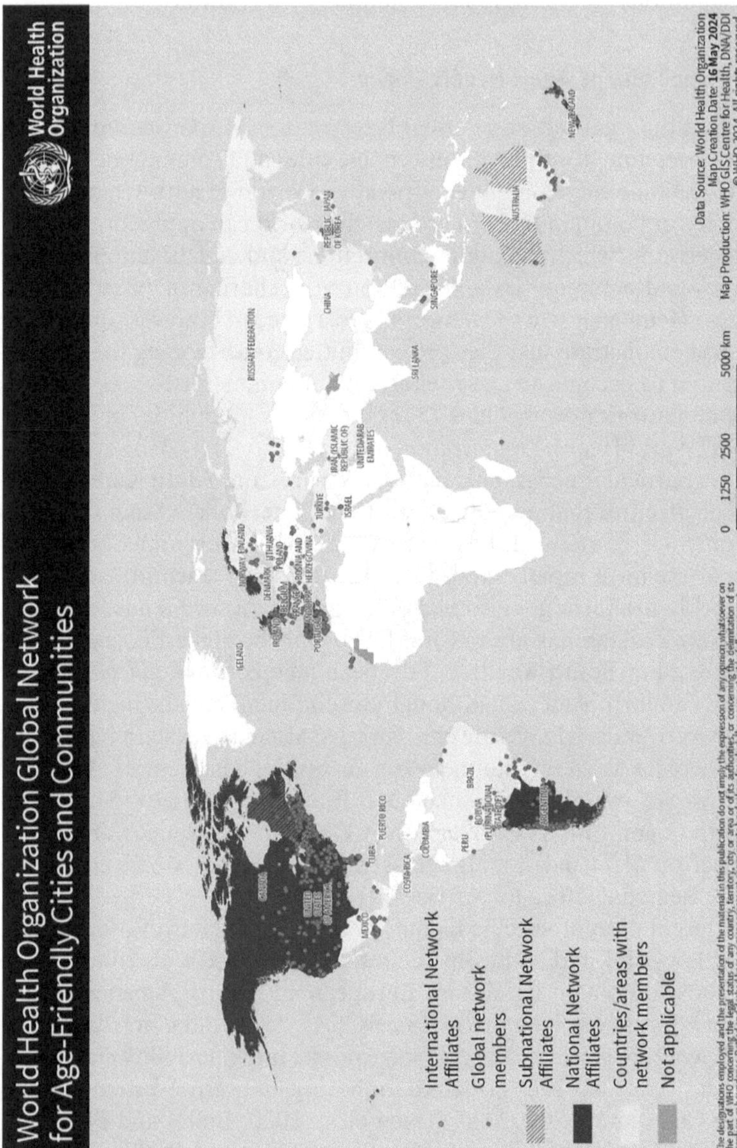

Figure 11.1 Map of the WHO Global Network for AFCC Members as of February 5, 2024. Credit: World Health Organization (2024). Source: https://extranet.who.int/agefriendlyworld/who-network/

of the global older population will be living in developing countries compared with 60% in 2005" (Steels 2015, 45). Steels' review of AFCC frameworks found that the frameworks as well as the studies they were based on were all from the developed world. "What is lacking is an ageing framework or model that can be used in both the developed and developing worlds" (Steels 2015, 48).

Critical literature on the geographies of development has long shown that imposing Western solutions or strategies onto non-Western cities and nations does not result in sustainable improvements to health, wealth, or quality of life (Ballard 2014). This is equally the case for the development of AFCC, and suggests that context-specific approaches are crucially important. Woolrych and colleagues, in a cross-national study that examined aging in place in cities across India, Brazil, and the United Kingdom, found that "the mechanisms through which place is constructed in old age vary across settings and are influenced significantly by urban transformations, changing mobilities in relation to place, cultural interpretations of age-friendliness, and environmental and psychosocial barriers to place participation" (Woolrych et al. 2021, 126). This suggests that a critical understanding of not only culture but also urban change, mobilities, and the social and environmental resources available to older people in a particular place is required in order to create AFCC in all regions of the globe.

Achieving this critical, nuanced perspective on age-friendliness requires attention on the part of researchers to the power structures within which cities, as economic centers, are embedded, and the increasing role of urban areas in the frameworks of everyday life around the world (e.g., Martynova, 2023). For example, the ways in which neocolonialism impacts countries such as Ghana may influence the changing role of the wage economy and the migration of younger people to urban areas, leaving a higher proportion of older people in rural and remote regions with fewer resources and services (Braimah, Rosenberg, and Kuuire 2022). In settler colonial societies, enduring colonial power structures interrupt intergenerational relationships within Indigenous nations: relationships that serve as foundational supports for older Indigenous people and provide continuity of Indigenous identity, history, and connections to the land (Lewis and Allen 2017; Wilson et al., 2010).

It is more useful to pay attention to specific urban and community contexts, rather than categorizing age-friendly capacity according to the arbitrary economic and cultural lines that characterize narratives of the "developed" and "developing" world. As the World Health Organization points out, "our physical and social environments, including family structure, social norms, and cultural traditions, influence people differently due to factors such as gender, ethnicity, sexual orientation, level of education, or disability" (World Health Organization 2023). The reason that factors such as ethnicity and sexual orientation influence people's lives in different ways has to do, at least in part, with the ways in which power structures play out within particular urban or community contexts. Attention to the role that power structures, such as historical and ongoing colonialism, play in the lives of older individuals in different contexts will lead to age-friendly policies that better reflect the challenges and opportunities that people encounter as they grow older. It will also position age-friendly policy to better support the diverse interests, strengths, and needs of older people both within and among cities.

The Role of Accessibility and the Physical Environment

In spite of the fact that age-friendly planning and policy include the social determinants of health and other aspects of the social and cultural environment, much AFCC literature emphasizes adaptation within the physical and built environment. Solutions that promote physical accessibility, mobility, transportation, and safe and affordable living environments for older individuals are often the priority. While these factors are certainly important in promoting an age-friendly urban environment and improving health and well-being in older age, scholars have called for a more integrated understanding of the person–environment relationship and a better understanding of the "places" of aging from the perspectives of older people themselves (Woolrych et al. 2021; Buffel and Phillipson 2012). Investigating the ways in which older individuals connect with place promotes understandings of community, home, and the spaces of health care that integrate physical accessibility with emotional and social landscapes. For example, barriers that impact an older person's use of public transportation might include physical difficulties with traveling to a stop and boarding a vehicle. However, accessible transportation that removes the physical barriers may follow a schedule that is not workable for the individual; an individual may feel unsafe while traveling alone; or an older person may have previously experienced discrimination on public transportation – all of which comprise barriers to using public transportation that have as much to do with social relationships to place and community as they do with the physical configuration of the transportation system and vehicles.

This suggests that an approach reminiscent of the WHO's Health in All Policies campaign (2014) would be effective in creating age-friendly cities, offering ways forward that add important dimensions of a holistic and critical understanding of the impacts of age-friendly cities policy on the everyday lives of older people around the world. Health in All Policies is intended to highlight the intersections among policy in different sectors and the ways in which sectors that may appear to have little to do with health – such as transportation – can have a huge impact on health. Health in All Policies recommends that various sectors work together with the overarching goal of creating a policy that enhances health and well-being. The AFCC policy inherently addresses multiple sectors at once and emphasizes intersectoral collaboration among physical and social environments (Steels 2015). Recognizing the synergy between age-friendly policy and planning, and Health in All Policies can usefully bolster the impact and reach of Health in All Policies while bringing attention to the need for healthy cities and health-promoting and -preserving urban infrastructure for all ages.

Although AFCC frameworks are intended to have global applicability, they are frequently based on research conducted in high-income countries, as discussed above (Fitzgerald and Caro 2014). Because of this ongoing geographical imbalance, AFCC literature frequently reflects a presumption that Western-style infrastructure is the best way to make a space age-friendly. Specific forms of public transportation, single-family housing, curated green spaces with walking paths, and the like are put forward as age-friendly solutions, but without acknowledging their cultural specificity. This can mean that age-friendly policy falls into a trap often critiqued in critical geographies of development of considering Western high-income countries as the gold standard which all other countries should aspire to emulate, when in fact there are many other ways to develop in an age-friendly manner that take local norms, infrastructures, and societal organization into account (World Health Organization 2023; Ballard 2014).

International frameworks such as Free, Prior, and Informed Consent (FPIC) (Schilling-Vacaflor and Flemmer 2020) can usefully be applied when researching AFCC. Scholars emphasize the need to have older people actively engaged in AFCC planning, if cities and communities are going to be age-friendly in ways that make a difference in the lives of older populations. Adopting the principles of FPIC is one way to ensure that older people in divergent cultural and geopolitical contexts have a chance to offer meaningful input. FPIC was first developed with respect to research with Indigenous communities and is designed to ensure community and participant input as well as oversight of any research with findings that can impact the communities and participants' lives. In order to deliver person-centered health and social services and ensure the engagement of older people in cities and communities, FPIC provides a useful guiding framework with which to approach collaboratively investigating age-friendliness with the people who stand to benefit directly from age-friendly policy and practice.

Also important are the impacts of cultural norms around aging when it comes to assessing and improving the age-friendliness of cities and communities. More attention needs to be paid to how older people are treated in old age and how this differs around the world. Existing cultural norms that value older people can be a good starting point in harnessing financial and human resources to create AFCC (World Health Organization 2023). In some parts of the world, for example, long-term care homes and institutional models of elder care are much less common than they are in the United States, and so "aging in place" may already be the most common method of caring for the elderly (Braimah, Rosenberg, and Kuuire 2022; Woo and Choi 2020). Age-friendly city and community models would benefit from taking non-Western cultures as examples and drawing some of their best practices from regions outside of the West, from Indigenous practices in settler-colonial nations, or from immigrant communities within both Western and non-Western contexts. Steels points out that "cities are a diverse mix of communities including different ethnic groups, social and private housing communities and socioeconomic groups"

(Steels 2015, 50). This makes it important to examine how both "cultural and socioeconomic influences can influence the success of age-friendly interventions in low- and middle-income countries" (Steels 2015, 49). In some cases, barriers to AFCC interventions may be related to a lack of resources; however, cultural contexts are equally important to consider. Ultimately, further research is needed to understand exactly which factors contribute the most to successful and impactful AFCC policy and practice. Scholars do not yet fully understand the causality and hierarchy of impacts on age-friendliness. "Are multi-stakeholder collaborations the key to age-friendly city success? Perhaps it is political involvement and financing; or investment in the social environment? And how does a city ensure that these interventions become sustainable?" (Steels 2015, 51). These are all questions to be pursued, with cultural norms and differences in mind, in order to understand the most effective ways to achieve age-friendliness.

Cultural norms related to aging are changing, in the West and beyond, and it is important to understand not only how they are changing, but what they are changing in response to (Braimah, Rosenberg, and Kuuire 2022). It is not enough to assume that changes in low- and middle-income countries are occurring due to the misconstrued model of development in which low-income countries develop to become more like high-income ones. More nuanced models and understandings of the impacts of urbanization and commodification need to be developed that prioritize the uniqueness of each country rather than their adherence to an assumed Western norm or ideal. For example, a strengths-based approach to aging in African countries could improve their engagement with the WHO global network for AFCC or highlight existing networks and knowledge communities in the region that detail best practices for health and well-being in later life.

Close ties to extended family, including between generations, and having established roles and responsibilities for elders within the family and community, have been shown to be important determinants of older people's health (Lewis and Allen 2017). In addition, the formation of immigrant communities with close ties to the country of origin as well as the community's new home has been shown to facilitate health and quality of life in older age as well as – at least in some cases – mitigate loneliness (Buffel, Phillipson, and Scharf 2012b). More work that takes the perspective of older individuals and is rooted in experiences of place will enhance understandings of the diversity of experiences of aging, and new approaches to aging well, across the globe.

The COVID-19 pandemic exacerbated many inequalities and highlighted housing, neighborhood, and healthcare inadequacies when it comes to the age-friendliness of urban areas (WHO Regional Office for Africa 2021; Finlay et al. 2021). It has taken a particular toll on older people, especially those living in congregated settings such as long-term care homes or multi-generational homes. According to the WHO, cities and communities with age-friendly policies in place are often better suited to respond in times of crisis, in an equitable way that is appropriate for all ages and abilities (World Health Organization 2023).

This needs to happen in all parts of the world, including in low- and middle-income countries, as well as in Indigenous nations, immigrant communities, and transnational communities, where cultural norms of aging and the elderly may contribute key insights to the global dialogue on AFCC.

Concluding Thoughts

If age-friendly policy is to gain traction in cities across an urbanizing world, not only do older adults and aging people across the lifespan need to be included as stakeholders and experts but also the power structures governing today's urban economic centers must be taken into account (Steels 2015; Buffel and Phillipson 2016). Geopolitical inequities as well as differing cultural norms around aging can limit the usefulness of age-friendly policy when it is applied – even flexibly – in non-Western regions of the world. Superficial adaptations of age-friendly policies from Western perspectives that are uncritically applied in non-Western contexts and emphasize infrastructure and the physical environment are unlikely to result in meaningful improvements for older populations around the world. Instead, recognizing and valuing expertise from varied cultural contexts on what it means to grow older and what roles older people can play in families, communities, and societies, can lead to more meaningful engagement as well as change.

Given that the framework for AFCC has been developed in an international context, research should follow that prioritize the perspectives of underrepresented communities and nations rather than maintaining a focus on countries that follow a preexisting norm or standard of what it means to be age-friendly. Taking the lead from critical urban studies, critical geographies of development, and the stated aims of the WHO and the UN Decade on Healthy Aging, researchers interested in promoting healthy urban environments for people of all ages should further develop place-based and experiential approaches to aging that respectfully and meaningfully involve older people themselves. This can draw on existing frameworks such as Free, Prior, and Informed Consent, which have been successfully used in research contexts in which inequities exist, to ensure that participants and their communities have not only input but also oversight over the development of age-friendly policy. Complementing function-oriented research in the physical environment and infrastructure with additional experiential, place-based research that explores how older people actually experience their environments will grow the body of work related to age-friendly cities and communities, give important consideration to cultural and geopolitical differences in understandings of age-friendliness, and focus on eliminating Western bias in order to reduce inequities on a global scale.

References

Ballard, Richard. 2014. "Geographies of Development III". *Progress in Human Geography* 39 (2). https://doi.org/10.1177/0309132514527031

Boudiny, Kim. 2013. "'Active Ageing': From Empty Rhetoric to Effective Policy Tool". *Ageing and Society* 33, pp. 1077–1098.

Braimah, Joseph Asumah, Mark W. Rosenberg, and Vincent Z. Kuuire. 2022. "'We are Left Behind to Suffer': Examining the Lived Experiences of Left-Behind Older Adults in the Upper West Region of Ghana". *African Geographical Review*. https://doi.org/10.1080/19376812.2022.2107548

Buffel, Tine, C. Phillipson, and T. Scharf. 2012a. "Ageing in Urban Environments: Developing 'Age-Friendly' Cities." https://doi.org/10.1177/0261018311430457csp.sagepub.com32(4)597-617

Buffel, Tine and Chris Phillipson. 2012. "Experiences of Place among Older Migrants Living in Inner-City Neighbourhoods in Belgium and England." *Diversité Urbaine* 11 (1). https://doi.org/10.7202/1007742ar

Buffel, Tine and Chris Phillipson. 2016. "Can Global Cities be 'Age-Friendly Cities'? Urban Development and Ageing Populations." *Cities* 55. https://doi.org/10.1016/j.cities.2016.03.016

Buffel, Tine, Chris Phillipson, and Thomas Scharf. 2012b. "*Experiences of Neighbourhood Exclusion and Inclusion among Older* People Living in Deprived Inner-City Areas in Belgium and England." *Ageing & Society*, *33*, pp. 89–109. https://doi.org/10.1017/S0144686X12000542

Finlay, Jessica M., Jasdeep S. Kler, Brendan Q. Oshea, Marisa R. Eastman, Yamani R. Vinson, and Lindsay C. Kobayashi. 2021. "Coping during the COVID-19 Pandemic: A Qualitative Study of Older Adults Across the United States." *Frontiers in Public Health* 9. https://doi.org/10.3389/fpubh.2021.643807

Fitzgerald, Kelly G. and Francis G. Caro. 2014. "An Overview of Age-Friendly Cities and Communities Around the World." *Journal of Aging & Social Policy* 26 (1–2). https://doi.org/10.1080/08959420.2014.860786

Herman, Lindsay, Ryan Walker, and Mark Rosenberg. 2020. "Age-Friendly Communities and Cultural Pluralism: Examining Saskatoon's Chinese-Canadian Social Enclave." *Canadian Journal on Aging / La Revue Canadienne Du Vieillissement* 40 (3). https://doi.org/10.1017/s0714980820000148

Lewis, Jordan P. and James Allen. 2017. "Alaska Native Elders in Recovery: Linkages between Indigenous Cultural Generativity and Sobriety to Promote Successful Aging." *Journal of Cross-Cultural Gerontology* 32 (2). https://doi.org/10.1007/s10823-017-9314-8

Martynova, Liliya. 2023. "The Concept of Active Longevity in Russia's Policy on Aging." *Journal of Aging & Social Policy*. https://doi.org/10.1080/08959420.2023.2265768

Nelson, Sarah E. and Mark W. Rosenberg. 2021. "Age-Friendly Cities and Older Indigenous People: An Exploratory Study in Prince George, Canada." *Canadian Journal on Aging / La Revue Canadienne Du Vieillissement* 41 (2). https://doi.org/10.1017/s071498082100009x

Paz, Amira, Israel Doron, and Aviad Tur-Sinai. 2018. "Gender, Aging, and the Economics of "Active Aging": Setting a New Research Agenda." *Journal of Women & Aging* 30 (3). https://doi.org/10.1080/08952841.2017.1295677

Schilling-Vacaflor, Almut and Riccarda Flemmer. 2020. "Mobilising Free, Prior and Informed Consent (Fpic) from Below: A Typology of Indigenous Peoples' Agency." *International Journal on Minority and Group Rights* 27 (2). https://doi.org/10.1163/15718115-02702008

Steels, Stephanie. 2015. "Key Characteristics of Age-Friendly Cities and Communities: A Review." *Cities* 47. https://doi.org/10.1016/j.cities.2015.02.004

Torku, Alex, Albert Ping Chuen Chan, and Esther Hiu Kwan Yung. 2020. "Age-Friendly Cities and Communities: A Review and Future Directions." *Ageing and Society* 41 (10). https://doi.org/10.1017/s0144686x20000239

UN Habitat. 2022. *World Cities Report*: UN Habitat.

WHO Regional Office for Africa. 2021. *Assessing the Impact of Covid-19 on Older People in the African Region, 2021*: WHO Regional Office for Africa.

Wilson, Kathi, Mark W. Rosenberg, and Sylvia Abonyi. 2010. "Aboriginal Peoples, Health and Healing Approaches: The Effects of Age and Place on Health." *Social Science & Medicine* 72 (3). https://doi.org/10.1016/j.socscimed.2010.09.022

Woo, Jung-Min and Moon Choi. 2020. "Why and how have Korean Cities Embraced the World Health Organization's Age-Friendly Cities and Communities Model?" *Journal of Aging & Social Policy* 34 (2). https://doi.org/10.1080/08959420.2019.1707057

Woolrych, Ryan, Judith Sixsmith, Jamuna Duvvuru, Adriana Portella, Mei Lan Fang, Deborah Menezes, James Henderson, Jenny Fisher, and Rebecca Lawthom. 2021. "Cross-National Perspectives on Aging and Place: Implications for Age-Friendly Cities and Communities." *The Gerontologist* 62 (1). https://doi.org/10.1093/geront/gnab170

World Health Organization. 2014. *Health in all Policies - Helsinki Statement. Framework for Country Action*: World Health Organization.

World Health Organization. 2023. *National Programmes for Age-Friendly Cities and Communities A Guide*: World Health Organization.

World Health Organization (2024). "*About the Global Network for Age-Friendly Cities and Communities.*" WHO Global Network for Age-Friendly Cities and Communities. https://extranet.who.int/agefriendlyworld/who-network/

Xiang, Liqun, Geoffrey Q. P. Shen, Yongtao Tan, and Xuan Liu. 2020. "Emerging Evolution Trends of Studies on Age-Friendly Cities and Communities: A Scientometric Review." *Ageing and Society* 41 (12). https://doi.org/10.1017/s0144686x20000562

12 On Equity and Designing Space for the Range of Human Ability

Ann Heylighen and Elizabeth Guffey

If you don't have access to physical spaces, (...) you don't have access to being able to build further your rights.

Team member, Studio Pacifica

Writing on design and disability has long been dominated by ideas of access and function but the belief that the built environment is connected to social equity is relatively recent. Although current attention to social justice has pushed a larger group of practitioners to position design and disability within growing calls for "diversity, equity, and inclusion," an explicit interest in equity in this connection is relatively new. The *Oxford English Dictionary* defines *equity* as "the quality of being equal or fair; fairness, impartiality; even-handed dealing," or as "what is fair and right."[1] Whereas equality treats everyone the same, equity recognizes that individuals differ in background, circumstances, and ability. Efforts to achieve equality can reinforce accessibility and usability as fundamental goals of human performance, but many approaches to disability in the built environment still overlook questions of fairness and social justice. Similarly, proponents of social emancipation and societal inclusion are not necessarily advocating for equity in design outcomes.

In this chapter, we explore equity in the built environment in relation to design for the range of human ability. We trace how equity features in design and disability discourse historically and connect insights from literature to experiences in contemporary architectural practice. To this end, we draw on interviews with founding principal Karen Braitmayer (KB) and three team members (T1–3) at Studio Pacifica,[2] a specialized consulting firm based in Seattle that "practices architecture from a place of supporting equity for people with disabilities."[3] For 30 years, the studio has been working with architects, civic planners, educators, and others to help create inclusive designs. Besides architectural expertise, the team relies on first-person disability experience, which makes it particularly well placed to help understand equity in the built environment in relation to universal/inclusive design practice.

DOI: 10.4324/9781032631110-13

Distinguishing Ease of Use, Independence, and Care

Some of the earliest research on design and disability was conducted in postwar Scandinavia, where ambitious social welfare policies pushed critics to reexamine disabled people's living conditions, economic support, employment, social services, and education; conceptualized as part of a public social security net, design research focused less on equitable outcomes and more on support to enable normalizing disabled people's lives. In design terms, this translated to a series of practical efforts to increase the ease of use and functionality of their homes. Special attention, for example, went to designing adaptive kitchens and living areas to be most usable for wheelchair users.[4]

The first conceptualization of equity in design and disability took shape as the barrier-free architecture movement began to crystallize in America in the 1950s. Much of this early work centered on issues of *equality*, with disabled people and their allies noting how social and physical barriers prevented them from living on equal terms with others. In design terms, the most extensive example of this approach appeared at the University of Illinois where rehabilitation therapist Timothy Nugent led a program for disabled students in the postwar years. At his behest, the University gradually made its campus more accessible to wheelchair-using students, radically reorganizing public and private space by widening narrow doors and replacing too small elevators and toilets, while also adding ramps and curb cuts on and around campus. Over time, advocates like Nugent began linking these simple but strategic architectural accommodations to broader and less tangible opportunities. In this formulation, a public building might theoretically be open to all, but steps at its entrance would prevent wheelchair users from entering; in effect, such architectural barriers could distinguish or discriminate against disabled people.[5] Although the term *equality* was never explicitly mentioned, the goal was for disabled people to live their lives "independently and without distinction."[6] In 1961 the American National Standards Institute published *American Standard Specifications for Making Buildings and Facilities Accessible to, and Usable by the Physically Handicapped* (ANSI 117.1), a relatively obscure six-page-long technical document that quantified this approach for the first time.

As the barrier-free architecture movement gained attention in the 1960s and 1970s, the notion that disabled people might best benefit from equal treatment in the built environment was not without challenges. A trained architect and polio survivor who often used a wheelchair himself, Selwyn Goldsmith revised his 1967 edition of *Designing for Disability* to provide a subtle critique of these ideas. Goldsmith's first 1963 edition was heavily influenced by ANSI 117.1, but his second edition was more extensive and influential; at the same time, it rethinks many of the assumptions underlying the American code. In this publication, Goldsmith conveyed an antipathy toward expressions of what he later characterized as an ideal of "self-help" and "independence" as reflected in the design of the built environment.[7] Arguing that his approach drew inspiration from Britain's new social welfare policies, Goldsmith believed that the American approach toward design and disability overemphasized the importance of personal autonomy.[8] Instead, he argued that dependence and interdependence were simply aspects of human existence.[9] Presenting his approach not as an argument for autonomy but rather an expression of care and equitable treatment, Goldsmith argued that disabled people could and should be better accommodated with well-marked and separate but parallel public facilities. This goal translated into design features current to this day, including the establishment of the *International Symbol of Access* (ISA) and wayfinding signage intended to guide disabled people to accessible entrances, toilets, and other accommodations, as well as the UK's provision of separate public toilet facilities outfitted for disabled people – and their caregivers.

Making Everyone Feel Welcome

Goldsmith's publications, designs for public toilet facilities, and insistence on the introduction of wayfinding signage are significant benchmarks for the implementation of more equitable forms of design for disabled people. But they also point to disabled people's ongoing efforts to translate theory into practice, highlighting the importance of real-world examples of equitable design used in daily life. As the difference between equality and equity – between "without distinction" and "separate but parallel" – can be quite abstract, the particulars may become clearer when we examine a specific practice in detail.

Ideas of care and equitable outcomes come to the fore in the practice of Studio Pacifica, which started in 1994 as a traditional architecture firm, in a partnership between Karen Braitmayer and a fellow architect. Today, Braitmayer explains,

> Studio Pacifica is *not* a traditional architecture firm. (…) We are a specialized consulting firm, and we focus on accessibility consulting, but (…) with a twist. There are many firms in the US that do what we do. I think what makes us stand out is that (…) because we have such a high percentage of staff that have lived experience, we (…) are pretty clear with our clients that we are there to help them not only to meet the [building] code requirements, but (…) look at ways to go beyond.

What Studio Pacifica helps their clients "to go beyond" is *equal access*, or *equality* (T1). When they think about a design, their priority is the human experience (T2). Rather than assuming that everyone experiences a design in the same way, they acknowledge that people's experiences may differ (T2):

> We definitely believe in […] design for a range of experiences. So we are always trying to think of the different disabilities and different ways that people will experience things with the range of abilities that there are.

Mindful of this range of experiences, Studio Pacifica helps their clients to ensure not only that everyone can access their building but also "that everyone feels *welcome*" (KB).

Their motivation to help their clients do so derives from personal lived experience. For Braitmayer and several team members, this includes experience with hearing loss and/or wheelchair use; one of them is a triple amputee and another is also vision impaired. For other team members, it includes experience with taking care of a child or parent in a wheelchair. Living daily with this range of abilities is what motivates Studio Pacifica to look at ways of making everyone feel welcome. Braitmayer explains:

> I can think of examples where I *don't* feel welcome. (…) when I walk into a small coffee shop or restaurant and all the tables are up here [shows table height] with stools, (…) I'm like okay, so you're pretty much telling me 'Don't come in.' I mean, you're not *saying* that, but you're kind of saying that. What we trying to help people see is: well, that means you're being exclusionary. And we don't want that.

Similarly, a team member in a wheelchair (T2) loves when Studio Pacifica looks at schools because of her own experiences at a young age:

> thinking about my own experiences growing up, using a wheelchair, what does social exclusion feel like, what does (…) the stigma of feeling like you're different, that you can't sit with your peers because there's nowhere (…) for a wheelchair to sit.

Another team member (T3) experienced transportation issues with her parents. Her mother over time needed to be in a wheelchair and she didn't have a vehicle that could take her places:

> And it was like wow, (…) here she is and she loves to go out places and see people and do things, but how to get her there? (…) So during the last summer she was with us, we rented an accessible van so I could take her out every night to enjoy the sunset. But it cost a ridiculous amount of money and I thought 'Oh my goodness,' I mean, she was very lucky that we were able to do that, but not everyone has the same opportunity for that, and when I think about the social justice part of that, that's not quite right to me.

Thanks to experiences like this, team members at Studio Pacifica are personally invested in helping their clients "go beyond" (T2):

> I will wake up at night and think 'Oh, did I check to make sure that there was door maneuver clearance?' (...) Things like that happen because I really do care, and I may never personally experience that building, but someone else is going to, and it can make their whole day a little better if they can get in those rooms.

Legal Rights and Voluntary Design Principles

While Studio Pacifica's motivation today resonates with the emphasis on care and equitable treatment Goldsmith expressed in the late 1960s, reference to ideas of equity and fairness for disabled people in the built environment is relatively recent. These ideas only became explicit with the rise of the Disability Rights Movement and the Independent Living Movement in the 1970s and 1980s. As the conversation shifted terms and goals, disabled people increasingly asserted themselves not as recipients of charity or patients within a medical or vocational program; instead, they demanded equal civil rights and opportunities as citizens. In the wake of these movements, significant legislation was enacted, including the *Americans with Disabilities Act* (ADA) (1990) and the UK *Disability Discrimination Act* (1995).

Congruent to the Disability Rights Movement, universal design was introduced in the 1980s as a direct reflection of ideas of *equitable use*. But universal design also took shape as disabled people and their allies found that barrier-free architecture and other accessibility codes were difficult to enforce. A driving force behind universal design, architect and wheelchair user Ron Mace himself experienced a lack of equity in schools and at work. Mace assisted in drafting government accessibility codes but found them at odds with unbending professional practices, disliked by practitioners troubled by layers of newer restrictions, and disconnected with lived disability experience.[10] In the early 1980s, Mace joined a circle of like-minded experts who began practicing human factors research while emphasizing the idea of "design for all people,"[11] a term that would ultimately be recast as *universal design*.[12]

These early efforts aimed to make the built environment "functional for all people," but issues of equity were only addressed directly in 1995, in a series of meetings under Ron Mace's leadership at North Carolina State University's Center for Universal Design. Some participants argued that "equitable use" was an anomalous category.[13] In fact, the other precepts focused on functionality and usability, whereas the idea of *equity* was deemed more theoretical and foundational. Moreover, the role and meaning of *equity* continue to be one of universal design's most debated features. While the other principles of universal design deal with "human performance" and "design strategy," in *Universal Design: Creating Inclusive Environments* architect Edward Steinfeld and researcher Jordana Maisel describe how universal design is tied to issues of social justice. Even they note, however, that *equity* is an imprecise term, sometimes being confused with the legal term "equal opportunity."[14]

More recently, some critics suggest that the 1995 definition of *equitable use* is simply insufficient when applied to the context of disability. Although scholar Aimi Hamraie agrees that the concept of universal design is rooted in design for disability, they note, "these guidelines do not appear to make an overarching ideological or value-based claim. Nor do any of the Principles mention disability, leaving unanswered the question of whom equity and flexibility are meant to benefit."[15] Geographer Ron Imrie has pushed still further, questioning how "equity" can figure in any construct that claims to be "universal." Imrie links the impulse to "universalize" human experience with Enlightenment positivism but argues that, by overlooking particularities, such frameworks lead to unequitable results.[16]

Differences in "Your Whatever"

Much like the concept of universal design, Studio Pacifica's work is rooted in lived disability experience, yet the team tends to a much broader user group than disabled people. While Braitmayer is known for her work with corporations, often in the technology sector, and public facilities – like the Seattle Public Library by OMA and LMN Architects or the Space Needle's renovation by Olson Kundig – Studio Pacifica works a lot on schools and low-income housing. Moreover, rather than *overlooking* particularities, the team draws designers' attention to them. They believe in universal design understood as design for a range of experiences, taking into account differences in ability, but also "your size, your age, your ethnicity, your religion, your whatever" (KB). And although Braitmayer is aware that her team is not ethnically diverse – "we're pretty much a [...] white persons group, [...] so that whole ethnic component [...], we're missing that" – she likes to think that they can also look through other lenses. When thinking about different experiences around lighting, for example, a team member (T1) points out:

> some people who (...) are extra sensitive to light (...) might want (...) less light switches or the ability to darken windows (...), and then some other people with low vision need it to be brighter so they can actually see what's going on, but then there are also people with no vision at all, not even light sensitivity, that (...) don't care at all. But there was also bumping up a bit on some cultural needs (...) a group of Muslim individuals that are like 'we need the ability block out our windows or else we'll have to wear our head scarfs hijabs inside constantly if there's the chance that the public can see in'.

Studio Pacifica is convinced that designers should take into account such particularities, yet they realize that meeting the building code requirements is not enough. "Ways to help people feel included," Braitmayer explains, "it's not well described in the code." At the same time, Studio Pacifica is not an enforcing body.

How then do they, as a consulting firm, help their clients ensure that everyone feels welcome in their buildings?

Part of the answer lies in the resources the team draws on, central to which are different kinds of experience. The first kind we already referred to is lived disability experience: people working at Studio Pacifica not only have different abilities themselves but their contacts in the disability community have different needs that the team sees secondhand and can also bring in (T1). Apart from underlying the team's motivation (KB, T2, and T3) and personal investment (T2), this lived disability experience offers a resource they draw on in multiple ways: as a sounding board when looking for solutions, but also as a set of examples to help clients understand why they want to change something (KB, T3, and T1). Instead of just saying "these are the requirements, and you must follow it," they try to help educate or inform their clients of the reasons (T3), and bringing their personal experiences together helps to provide detail in doing so (T3 and T1). Moreover, when being physically present during site visits, they "lend their personal experience" to actually show their clients rather than talking about hypotheticals: "They're like: 'but this is the code.' I'd be like: 'Yeah, it's the code, but just look at how it works, or doesn't work'" (T1).

The building code brings us to a second kind of experience Studio Pacifica relies on, which is Braitmayer's involvement in developing and improving guidelines and standards. For 13 years she served as a Presidential Appointee to the US Access Board, a committee that develops guidelines and standards for the built environment under the ADA and other laws. Moreover, she is actively involved in updating the abovementioned *ANSI 117*, currently called the *American National Standard for Accessible and Usable Buildings and Facilities*. While Studio Pacifica wants their clients to go beyond meeting building code requirements, Braitmayer's experience in code development gives the team a nuanced understanding of the rationale behind these requirements:

> because Karen knows the history of *why* something is in a code usually, she can explain what they were actually intending, and sometimes what they were intending doesn't exactly align with what we thought they were meaning.
>
> (T3)

This nuanced understanding in turn helps Studio Pacifica explain to their clients why they recommend "to go beyond" meeting the building code. Mindful that they are not an enforcing body, they seek analogous language in the code to justify their recommendations.

Emphasizing Belonging, Reducing Stigma, and Recognizing Who Benefits Most

Even as universal design's appeal to *equitable use* has been debated, its explicit call for equity differentiates it from more recent developments, including approaches to inclusive design and design for all. First introduced by design researcher Roger Coleman in 1994, the term "Inclusive Design" may be distinguished from universal design precisely because of such particularities. Coleman argues that universal design's rhetoric and practice position it as a discrete category of design; establishing separate precepts and principles like equitable use separates universal design from mainstream methods and practices. Furthermore, at least in Coleman's initial definition, Inclusive Design focused primarily on the needs of older people. Above all, he argued, "by taking account of this in the design process, products, services and environments can be improved for the majority of customers in ways that are not associated with negative perceptions of age or disability."[17] Emphasizing social belonging, Coleman and later proponents like engineering designers Simeon Keates and John Clarkson have argued that Inclusive Design be positioned in opposition to the concept of design exclusion.

At the same time, questions of equity and fairness have been raised here too. Insisting that architectural professionals have misunderstood "inclusive design," architect Matteo Zallio has joined Clarkson in arguing for an expanded notion of *inclusion* to encompass social justice issues as well. Arguing that "inclusion is not merely a matter of making a design that works well for people with disabilities," Zallio and Clarkson have called for an "extended definition" of the term that "also includes understanding how people behave, how they socialise, how they live and how they access the space."[18] In line with the experiences of Studio Pacifica's team members, other critics point toward the role that stigma plays in understanding such issues. Designer Joshua Halstead specifically links equity with a reduction of stigma, arguing that this – along with increased access and participation – can lead holistically to more equitable design outcomes.[19]

Notwithstanding such arguments, scholars Matteo Bianchin and Ann Heylighen have identified the "paradox of inclusive design" behind efforts to achieve equity: "To what extent is it possible," they ask, "to design a product, space or service that, at the same time, allows for equitable use by everyone and respects the diversity in people's capacities?"[20] Drawing upon philosopher John Rawls' writings,[21] Bianchin and Heylighen suggest that equity can best be achieved if designers ultimately "accept some difference in usability provided that the solution maximizes the results for the worst offs in general."[22] Similarly, scholar Sasha Costanza-Chock has argued for the place of disability within the larger "Design justice" movement and positions equity as pivotal to these efforts. Focusing on justice-based values like equity, Costanza-Chock argues, "challenges the underlying assumption that it is possible to design for all people. Instead, we must always recognize the specificity of which kinds of users will benefit most."[23]

Offering Choice

Besides the fact that people's experiences differ, Studio Pacifica acknowledges that something which benefits one community may hamper another. For instance Braitmayer explains,

> the low vision and blind community [...] are strongly encouraging using tactile material on the walking surfaces, in particular like in large plazas [...] that help people get from point A to point B. Well, now you have a bumpy thing on the ground surface. And [...] oftentimes that bumpy material goes to the front door, and people who are wheelchair users are like 'I don't like that!'

In case of such contradictory needs, Studio Pacifica looks for "the happy medium" (KB), for example, by making mock-ups of tactile guidance on walking surfaces and testing materials on the ground with wheelchair users and vision-impaired people. Or they try to think about things differently (T3), like Braitmayer did for the redesign of a bar (Figure 12.1): at first sight, the height of a bar required for someone seated on a stool does not work for someone in a wheelchair or of short stature, and vice versa. Braitmayer realized, however, that it can work when floor levels are different.

More generally, while universal design is criticized for its alleged *one-size-fits-all* solutions, Studio Pacifica advocates offering choice: "In conditions where different bodies are using a similar feature," say a paper towel dispenser, "why does that feature always have to be the same height? (...) architects like everything lined up, visually. But that doesn't necessarily serve the community well" (KB). In an existing building, the team may recommend adding an extra

Figure 12.1 Deep Dive, a bar in Seattle. Graham Baba designed it in collaboration with Studio Pacifica. Credit: Suzi Pratt.

feature at a different height instead of removing what is there already. In the case of a new building, offering choice may be providing a base level plus room to add extra accessibility later if needed; for instance, "all bathrooms have blocking in the walls to add grab bars if someone needs them, but it doesn't necessarily start there" (T1).

A Reasonable Answer

Even as conversations about design and disability flourish, there has been a blurring of various approaches. Moreover, a broader interest in addressing social justice questions more directly has pushed the design profession to re-evaluate its efforts to promote "diversity, equity and inclusion" (DEI). Understanding that disability is only sometimes positioned within these efforts, there have been calls to add the question of "accessibility."[24] At the same time, recent ideas of inclusive environments expand toward "all groups that have been excluded from full participation" in society,[25] extending attention in design from disability to those marginalized by culture, religion, gender, language, LGBTQ+ identity, and other intersectional and individual identities. In this way, there has been increasing interest in applying disability theory – and ideas of equity – to a range of building practices.[26] Along these lines architect Joel Sanders, for example, has shaped the *Stalled!* project to design unisex and accessible public toilets. The project applies ideas of accessibility to address civil liberties for the "need to create safe, sustainable and inclusive public restrooms for everyone regardless of age, gender, race, religion and disability." [27]

When implementing such projects in practice, architects always operate in a condition of moderate scarcity of resources: their agency is constrained by resources being insufficient to satisfy all possible needs. Having worked as an architect before, Braitmayer has been the one trying to build on budget and on time for a client. Besides lived disability experience and experience in code development, she and other team members at Studio Pacifica have experience in different aspects of architectural practice, from design to project management. This experience enables them to empathize with the architects they collaborate with: they value the architects' work and realize that there are so many codes and things that architects are dealing with in addition to accessibility issues (T3). Moreover, it motivates and equips them to support the architects' design, trying to keep their design idea and get it to work:

> we've received (…) plans (…) with triangular toilet rooms. Okay, so, (…) *very hard* to fit the code into that. But we (…) totally get that you want that. (…) So we're (…) trying to work with them, to try to keep their design idea *and* get it to work in the code. And (…) it's fun, I mean, that part is fun, right. Because [the architects] always push the limits. (…) They do beautiful, beautiful work. And so it's an honor to be involved.
>
> (KB)

Drawing on the combination of their experiences enables Studio Pacifica to help clients "build the best project they can" (KB). It enables them to provide clients with trusted information; so they feel like they're getting value for their money: "Like, we're telling you the right thing and it's a *reasonable* answer, not like a totally wacky, out there, (…) you know, ultra-conservative answer" (KB).

Being reasonable includes acknowledging that there is not always a good answer to ensure that everyone feels welcome. A case in point is a series of elementary schools Studio Pacifica is working on through a public school district. The standard elementary school has a big open room for lunchtime and gatherings, and adjacent to that is a space that is raised about 24 inches and that could be a stage. Now it is designed with two sets of stairs up to that space, and the wheelchair accessible route is through a ramp that starts down the hall. Both the building code and the ADA say that you must have a direct connection from a seating area to the stage.

> And, so there's a debate: is asking a kid who's a chair user to go down the hall and come up the ramp the same as the kid who gets to go up three steps up right there? And we would say: 'No, uhm, it's not the same.' (…) But there's a piece of me that feels like, you know, 'It's already built, now what?' (…) So, that's the struggle that we have. And there isn't always a good answer.

Indeed, whereas ideas of access and independence have helped shape thinking on design and disability for several decades, problems shaping fair and equitable solutions in the built environment have only recently begun to be addressed. While there may be no perfect solutions to such problems, firms like Studio Pacifica still strive for greater equity in design. The ultimate goal may be not to solve all problems at all times but rather to take care "that everyone feels *welcome* (KB)."

Acknowledgments

This work was supported by the Francqui Foundation in the form of Ann's Research Professorship. She is deeply grateful to Studio Pacifica for their hospitality, their patience, and the generosity with which they shared their time, views, and experiences; to the colleagues who took over her teaching tasks during her Research Professorship; and to Daniel S. Friedman, Koen Coomans, Natalia Pérez Liebergesell, and Sam Michiels, for the invaluable support, discussions, and feedback.

Notes

1 OED, "equity," accessed February 15, 2024. https://www.oed.com/search/dictionary/?scope=HistoricalThesaurus&q=equity.
2 The first author conducted semi-structured interviews with Karen Braitmayer and three team members who prefer to remain anonymous and are referred to as T1, T2,

and T3. To this end, she visited Studio Pacifica's office in Seattle, with two interviewees participating online. All interviewees gave written and oral consent after reading a form describing the study's ethical and confidentiality provisions, as approved by the Social and Societal Ethics Committee at KU Leuven. Interviews took between 1 and 1, 5 hours each; they were in English, audio-recorded, transcribed, and complemented with information on Studio Pacifica's website and other documents about their work. The first author analyzed the materials using a combination of inductive and theory-driven coding.

3 "Now hiring!", Studio Pacifica, accessed March 8, 2024, https://studiopacificaseattle.com/jobs/.

4 See for example Bengt Erik Eriksson and Rolf Törnqvist, eds., *Likhet och särart. Handikapphistoria i Norden* [Equality and difference. Disability history in the Nordic countries] (Nynäshamn: Handikapphistoriska föreningen, 1995); Greta M. Cederstam, ed., *Vägen till människovärde. Några drag ur nordisk handikapphistoria åren 1945–1985* [The road to human dignity. Some features from Nordic disability history in 1945–1985] (Vällingby: Nordiska nämnden för handikappfrågor, 1990).

5 See Elizabeth Guffey, *Designing Disability: Symbols, Spaces and Society* (London: Bloomsbury, 2017); Bess Williamson, *Accessible America* (New York: New York University Press, 2020).

6 Timothy J. Nugent, *Design of Buildings to Permit Their Use by the Physically Handicapped* (Washington, D.C.: National Academy of Sciences, National Research Council, fall 1960), p. 53.

7 Selwyn Goldsmith, "The Ideology of Designing for the Disabled," Proceedings of the. Fourteenth International Conference of the Environmental Design (Lincoln, Neb., January 1983), pp. 198–214.

8 See Elizabeth Guffey, "Selwyn Goldsmith's designing for the disabled, 2nd ed. (1967): Flawed, dated, and disavowed, yet a classic with enduring value," *She Ji: The Journal of Design, Economics, and Innovation* 6, no. 4 (2020): pp. 439–454.

9 At this time, Goldsmith was deeply influenced by the writings of social psychologist Beatrice Wright, who advocated for the social and psychological – as well as the physical – well-being of disabled people. However, by the early 1980s Goldsmith himself began to pivot, and ultimately disavowed the second and third editions of his book. In a fourth edition of his text (*Designing For the Disabled: The New Paradigm* (New. York: Routledge, 1997)), he repudiated the ideas he had promoted in his well-known second and third editions and argued that the American approach had been right all along.

10 Aimi Hamraie, *Building Access* (Minneapolis: University of Minnesota, 2017), p. 208.

11 Elaine Ostroff and Daniel Iacofano, "Teaching Design for All People: The State of the Art," Design Faculty Seminar, Adaptive Environments Center, Boston, Mass., April 1–3, 1982, Elaine Ostroff Universal Design Papers, Archives Center, National Museum of American History, Smithsonian Institution.

12 Ronald Mace, "Universal Design: Barrier-Free Environments for Everyone," *Designers West* 33.1 (1985): pp. 147–157.

13 Molly Follette Story, "The Principles of Universal Design," *The Universal Design Handbook* (New York: McGraw Hill, 2001), p. 10.3.

14 Edward Steinfeld and Jordana Maisel, *Universal Design: Creating Inclusive Environments* (Hoboken, NJ: Wiley), 88.

15 Aimi Hamraie, "Designing Collective Access: A feminist disability theory of universal design," *Disability, Space, Architecture: A Reader* New York: Routledge, 2016, p. 81.

16 Rob Imrie, "Universalism, Universal Design and Equitable Access to the Built Environment," *Disability and rehabilitation* 34 (2011): p. 876.

17 Roger Coleman, "The Case for Inclusive Design—An Overview," *Proceedings of 12th Triennial Congress, International Ergonomics Association and the Human Factors Association* (1994), p. 251.

18 Matteo Zallio and John Clarkson, “Inclusion, diversity, equity and accessibility in the built environment: A study of architectural design practice,” *Building and Environment*, 206 (2021) https://www.sciencedirect.com/science/article/pii/S0360132321007496.
19 Joshua Halstead, “Equations for Reducing Disability Stigma through Design Equity,” *After Universal Design*, ed. Elizabeth Guffey (London: Bloomsbury, 2023), p. 144.
20 Matteo Bianchin and Ann Heylighen, “Just Design,” *Design Studies* 54 (2018): pp. 1–22.
21 See John Rawls, *A Theory of Justice* (Cambridge, Mass: Harvard University Press, 1971); *Political Liberalism* (New York: Columbia University Press, 1993); “Justice as Fairness,” *Philosophy & Public Affairs* 14, no. 3 (1985): pp. 223–251.
22 Bianchin and Heylighen, “Just Design,” 19.
23 Sasha Costanza-Chock, *Design Justice: Community-Led Practices to Build the Worlds We Need* (Cambridge, Mass: MIT Press, 2020), p. 66.
24 Zallio & John Clarkson, “Inclusion, diversity, equity and accessibility in the built environment”.
25 Steinfeld & Maisel, *Universal Design*, 41.
26 See, for example, Victoria Lanteigne, Traci Rose Rider, and Peter Stratton, “Evolving Design Pedagogies: Broadening Universal Design for Social Justice,” *Enquiry The ARCC Journal for Architectural Research* 19, no. 1 (September 5, 2022): pp. 8–23, https://doi.org/10.17831/enqarcc.v19i1.1135.
27 “Stalled!,” Stalled!, accessed January 23, 2024, https://www.stalled.online.

Bibliography

American National Standards Institute et al. *ANSI 117.1, American Standard Specifications for Making Buildings and Facilities Accessible to, and Usable by the Physically Handicapped.* New York: American Standards Association, 1961.

Bianchin, Matteo and Ann Heylighen. “Just Design.” *Design Studies* 54 (2018): 1–22.

Cederstam, Greta M., ed. *Vägen till människovärde. Några drag ur nordisk handikapphistoria åren 1945-1985 [The Road to Human Dignity. Some Features from Nordic Disability History in 1945-1985].* Vällingby: Nordiska nämnden för handikappfrågor, 1990.

Coleman, Roger. “The Case for Inclusive Design—An Overview.” *Proceedings of 12th Triennial Congress, International Ergonomics Association and the Human Factors Association*. Toronto 1994.

Costanza-Chock, Sasha. *Design Justice: Community-Led Practices to Build the Worlds We Need.* Cambridge, Mass: MIT Press, 2020.

D’Souza, Newton. “Is Universal Design a Theory?” *Designing a More Inclusive World*, edited by Simeon Keates, John Clarkson, Patrick Langdon, and Peter Robinson. London: Springer, 2004.

Eriksson, Bengt Erik, and Rolf Törnqvist, eds. *Likhet och särart. Handikapphistoria i Norden* [*Equality and Difference. Disability History in the Nordic Countries*]. Nynäshamn: Handikapphistoriska föreningen, 1995.

Goldsmith, Selwyn. “The Ideology of Designing for the Disabled.” *Proceedings of the Fourteenth International Conference of the Environmental Design. Lincoln, Neb.*, January 1983, 198–214.

Goldsmith, Selwyn. *Designing For the Disabled: The New Paradigm.* New York: Routledge, 1997.

Guffey, Elizabeth. *Designing Disability: Symbols, Spaces and Society*. London: Bloomsbury, 2017.

Elizabeth Guffey, "Selwyn Goldsmith's Designing for the Disabled, 2nd ed. (1967): Flawed, Dated, and Disavowed, Yet a Classic with Enduring Value," *She Ji: The Journal of Design, Economics, and Innovation*, 6, no. 4, 2020, pp. 439–454.

Halstead, Joshua. "Equations for Reducing Disability Stigma through Design Equity." *After Universal Design*, edited by Elizabeth Guffey. London: Bloomsbury, 2023.

Hamraie, Aimi. "Review of Universal Design as a Rehabilitation Strategy (Sanford) and Universal Design: Creating Inclusive Environments (Steinfeld & Maisel)." *Design and Culture* 5, no. 2 (2013): 268–271.

Hamraie, Aimi. "Designing Collective Access: A Feminist Disability Theory of Universal Design." *Disability, Space, Architecture: A Reader*, edited by Jos Boys. New York: Routledge, 2016.

Hamraie, Aimi. *Building Access*. Minneapolis: University of Minnesota, 2017.

Imrie, Rob. "Universalism, Universal Design and Equitable Access to the Built Environment." *Disability and Rehabilitation* 34, no.10 (2011): 873–882.

Keates, Simeon and John Clarkson. *Countering Design Exclusion*. London: Springer, 2004.

Lanteigne, Victoria, Traci Rose Rider, and Peter Stratton. "Evolving Design Pedagogies: Broadening Universal Design for Social Justice." *Enquiry The ARCC Journal for Architectural Research* 19, no. 1 (September 5, 2022): 8–23, https://doi.org/10.17831/enqarcc.v19i1.1135

Mace, Ronald. "Universal Design: Barrier-Free Environments for Everyone." *Designers West* 33, no. 1 (1985): 147–157.

Nugent, Timothy J. "*Design of Buildings to Permit Their Use by the Physically Handicapped.*" National Academy of Sciences, National Research Council, Washington, D.C., fall 1960.

Ostroff, Elaine, and Daniel Iacofano. "Teaching Design for All People: The State of the Art." Design Faculty Seminar, Adaptive Environments Center, Boston, Mass., April 1–3, 1982, *Elaine Ostroff Universal Design Papers, Archives Center, National Museum of American History, Smithsonian Institution.*

Oxford English Dictionary. "Equity." Accessed February 15, 2024. https://www.oed.com/search/dictionary/?scope=HistoricalThesaurus&q=equity

Rawls, John. *A Theory of Justice*, Cambridge, Mass: Harvard University Press, 1971.

Rawls, John. "Justice as Fairnes." *Philosophy & Public Affairs* 14, no. 3 (1985): 223–251.

Rawls, John. *Political Liberalism* New York: Columbia University Press, 1993.

Sandu, Jim. "The Rhinoceros Syndrome." *The Universal Design Handbook*, 2nd ed., edited by Wolfgang F. E. Preiser and Korydon H. Smith. Hoboken: McGraw Hill, 2011.

Steinfeld, Edward, and Jordana Maisel. *Universal Design: Creating Inclusive Environments*. Hoboken, NJ: Wiley, 2012.

Story, Molly Follette. "The Principles of Universal Design." *The Universal Design Handbook*, 1st ed., edited by Wolfgang F. E. Preiser and Elaine Ostroff. New York: McGraw Hill, 2001, 10.3–10.19.

Studio Pacifica. "Now hiring." Accessed March 8, 2024. https://studiopacificaseattle.com/jobs/

Williamson, Bess. *Accessible America*. New York: New York University Press, 2020.

Zallio, Matteo, and John Clarkson. "Inclusion, Diversity, Equity and Accessibility in the Built Environment: A Study of Architectural Design Practice." *Building and Environment* 206 (2021), https://www.sciencedirect.com/science/article/pii/S0360132321007496

13 Auditory Ableism

Exploring Aural Diversity, Equity and Belonging in the Built Environment

William Renel

Introduction: Multimodality, Ableism and the Built Environment

As humans our experiences of the urban world are multimodal with different sensory modes (hearing, touch, vision, smell, temperature etc.) contributing to our overall social, political and embodied experiences of the built environment. Adopting different forms of analysis has enabled the built environment to be examined and scrutinised from different perspectives. Multimodal analysis suggests that the built environment influences human behaviours and feelings through different sensory channels (Ravelli and McMurtie, 2016). Therefore, the analysis that focuses on an individual mode (e.g. vision, physical form or structure) is limited to a partial understanding of the complexities and realities of what it means to be human within the built world. Existing discussions of power in the built environment (Dovey, 1999; Soja, 2011) contend that architecture and design serve as mediators of the cultural and social practices of power. Rieger (2023) expands on this, discussing the social, cultural and structural practices, which are manifested in the built environment and mediated spatially, to create and maintain experiences of ableism. Ableism is a system of oppression comprised of beliefs, values and practices that present disability as a devalued state of being (Ellis et al., 2019). A Sound Studies approach to the analysis of the built environment centres sound and hearing and considers "what sound does in the human world, and what humans do in the sonic world" (Sterne, 2012: 02). This perspective creates the opportunity to challenge ocular binaries and "provide a means for spaces, people, and objects to resound and articulate that the impossibility of closing an earlid is a possibility for more socially just, ecological methodologies" (Daza and Gershon, 2015: 01). Analysis of the built environment from a Sound Studies perspective has given rise to the notion of aural architecture – described as the composite of surfaces, objects and geometries in an environment and how the properties of such things in space are experienced by listening (Blesser and Salter, 2019). As well as providing acoustic cues, interpreted through objects and surfaces, aural architecture influences our feelings, behaviours, moods and associations. However, the divergent realities of human hearing and the emerging concept of "aural diversity" have so far been underutilised in discourse and discussion

DOI: 10.4324/9781032631110-14

surrounding aural architecture, ableism and spatial inequality in the built environment.

Aural Diversity

The term aural diversity recognises the multitude of factors that position hearing in a constant state of flux (Renel, 2018). Drever and Hugill (2023) suggest that aural diversity is not a set of textbook definitions stipulating metrics, but rather an acknowledgement of the complexities of lived and embodied hearing experiences in all their diversity and fluctuation. Drever (2017) provides a useful description of aural diversity as the "actual variety of (often less than ideal) hearing that we experience throughout a normal day and through-out our lives". Factors that influence the aural diversity of the population include, but are not limited to, age, neurodiversity, hearing loss, auditory conditions (e.g. tinnitus), nonauditory conditions (e.g. hypertension), sight loss and temporary factors (e.g. ear infections). The aural diversity paradigm, formalised by the Aural Diversity Network in the UK (auraldiversity.org 2023), highlights the multiplicity of ways that sound can affect individuals via the divergent hearing profiles present within the population.

The Auditory Normate

In previous research, the author has aimed to broaden the academic conversation about how the aural diversity paradigm shapes and is shaped by design. A central tenant of work in this area is the construction of a new term: the "auditory normate" – an idealised sonic citizen around which the contemporary world is built (Renel, 2023). This term enables us to consider how the design and management of objects, environments and services produce auditory hierarchies where certain forms of hearing, listening and communication are given priority over others. This is in short, a description of auditory ableism. Examples of auditory ableism propagated through design are automated phone services, environments governed by auditory rules (e.g. libraries and galleries) and the placement of public information speakers, all of which are created with a normative perspective in mind. Auditory ableism is also embedded in acoustics legislation and policy. For example, the acoustic standard for normal equal-loudness-level contour (BS ISO 226, 2003), a widely utilised document in urban design that dictates the acceptable level of ambient noise within a built environment, is based on an "otologically normal person". The standard describes this as someone in a

> normal state of health who is free from all signs or symptoms of ear disease and from obstructing wax in the ear canals, and who has no history of undue exposure to noise, exposure to potentially ototoxic drugs or familial hearing loss.
>
> (BS ISO 226, 2003)

The standard also stipulates an age range of 18–25 years – the period of life when human hearing is often strongest and healthiest. LaBelle (2019, 190–191) notes that the practice of

> acoustics is shaped by the normative patterns that often define spaces, contributing to what can be heard and where, who may speak or not, and what types of behavior can enter into the time signatures of an environment and by whom.

Auditory ableism in design can be theorised as a sociopolitical sounding of oppression or what Kanngieser (2011) describes as an acoustic politics of the body and voice. Auditory ableism also emphasises what Stoever (2016) defines as the "listening ear" – a socially constructed ideological system producing but also regulating normative cultural ideas about sound.

Interviews and Soundwalks

Building on the above, this chapter centres on the experiences of disabled and/or neurodivergent people through the analysis of two data sources. The first is a set of semi-structured interviews and the second is a series of recorded explorative soundwalks in urban environments across the UK. The interviews were undertaken using video conferencing software, enabling participants to take part from a location of their choosing and at a date and time that suited them. The interviews lasted approximately one hour, were recorded as audio and then transcribed into text. In total, five interviews were undertaken, and all participants were paid for their time. The purpose of the interviews was to enable disabled and neurodivergent participants to share their experiences of equity, sound and the built environment in their own words and on their own terms. This approach to narrative interviewing is informed by the method of counter-storytelling (Sitter and Grittner, 2021), which has its roots in critical race theory (Delgado, 1989; Solórzano and Yosso, 2002) and has subsequently been championed by critical disability studies as an inclusive and accessible research practice (Rieger, 2023).

The explorative soundwalks enabled data to be collected in an experiential way. Soundwalking is a method of attentive listening to the sounds within an environment. It was pioneered within the field of Acoustic Ecology, notably by Hildegard Westerkamp who describes soundwalking as "any excursion whose main purpose is listening to the environment … the intention is always to rediscover and reactivate our sense of hearing within a specific context" (Westerkamp, 1974). During each of the soundwalks in this study, a disabled and/or neurodivergent participant was invited to discuss the role of aural diversity and soundscape design in shaping feelings of equity within the urban environment. The soundwalks were recorded using binaural microphones; in-ear microphones worn by the participants that capture the conversation alongside a 3D image of the evolving sound environments. In total, six recorded soundwalks were undertaken, each lasted approximately one hour and all participants were paid for their time. The soundwalks included multiple locations such as high streets, cafes, museums, theatres and galleries. The soundwalks were undertaken in cities across the UK including London, Glasgow and Sheffield. Each of the total 11 participants who contributed to the study was asked to provide a brief description of themselves to be included in this chapter.

1. Participant A is 52, identifies as male, white British and lives in London. He is visually impaired.
2. Participant B is 36, identifies as male, black African and lives in London. He is visually impaired with Nystagmus.
3. Participant C is 49, identifies as non-binary, white British and lives in Sheffield. They are neurodivergent and experience voice hearing.
4. Participant D is 30, identifies as female, white British and lives in Cambridge. She is deaf.
5. Participant E is 25, identifies as female, white British and lives in Brighton. She is neurodivergent and has Meniere's disease.
6. Participant F is 28, identifies as female, mixed British and lives in Rotherham. She is neurodivergent and has chronic fatigue.
7. Participant G is 35, identifies as female, mixed British and lives in Bristol. She is neurodivergent and has hearing loss.
8. Participant H is 60, identifies as male, mixed British and lives in London. He is autistic and experiences hyperacusis and misophonia.
9. Participant I is 23, identifies as female, white British and lives in Glasgow. She is neurodivergent and is an ambulatory wheelchair user.
10. Participant J is 44, identifies as female, white British and lives in Sheffield. She is a wheelchair user with hearing loss in one ear.
11. Participant L is 28, identifies as non-binary, Asian and lives in Birmingham. They are neurodivergent and have epilepsy and hyperacusis.

Throughout the following sections of this chapter, the participants are quoted directly.

Analysis

Content analysis is a research technique that is used to examine data (most often written text or media such as photographs and videos) to ascertain and interpret the meaning and themes that are contained in the data (Neuman, 2011). When content analysis is undertaken within the field of disability studies, the intention is often to extract counter-narratives – examples from within the data that disrupt or critique a dominant position, such as ableism (Berger and Lorenz, 2017). Therefore, using content analysis to examine the data within this chapter has two primary intentions. First, to find themes within the data that evidence how sound and soundscape design in the UK contribute to experiences of ableism. Second, to extract counter-narratives that critique the perspective of the auditory normate.

Theme 1: Sound and Belonging

Zouras (2020, 01) suggests that although

> designers, urban planners, and decision-makers who shape the environment know that the space in which people occupy is essential to their overall welfare, uncertainty lies in determining which specific design elements, components, and qualities contribute to perceived feelings of belonging.

Reed and Bohr (2021) describe the significance that local built environments have on experiences of social belonging. Yuval-Davis (2006) develops an analytical framework for the study of the politics of belonging, exploring both historical and contemporary perspectives on the interplay between social exclusion and feelings of belonging. LaBelle (2019, 177) brings an auditory positionality to discussions of the built environment and belonging contending that sound and acoustics are "central to engaging a cultural politics of hearing and belonging … the propagation of a sound wave is thus underscored as a tensed, energetic volume around which sharing and belonging, rapture, and fragmentation are held". It is evident that there is not a single framework through which belonging is understood within existing research and discourse on the built environment. However, a scoping study by Raines et al. (2023) suggests that within the field of disability studies, common characteristics that articulate one's sense of belonging include being present, welcomed, supported, accepted, known and loved. The scoping study also highlights the lack of input from learning-disabled and neurodivergent people in existing research concerned with belonging.

Each of the interviews and soundwalks within this study started with a discussion of how sound and hearing shape feelings of belonging in the built environment and how this affects the participants' feelings of equity. Belonging was framed in short as an affinity for a place or experience and equity as the feeling that something is fair and equal. In describing a residential urban environment in North-East London where he lives, Participant A commented:

> I certainly feel I belong to the road I live on. I really do feel like I have a different relationship to this road that obviously wasn't there when I first moved in … local sounds (like kids playing in the school playground at lunchtime) give me a sense of confidence, a sense of presence, that I'm physically aware of what's going on in my neighbourhood. And therefore, I'm up to date and socially connected. The sounds locally make me feel that things are equal, even though I am blind. Because I use sound to know what's going on.
>
> (Participant A)

Participant B discussed the significance of sound in shaping their feelings of belonging in the built environment as someone with a visual impairment:

> Sound out of all the senses probably has the biggest impact on whether I feel like an urban environment is an environment that I belong to. Vision is very different. You can turn away from something that you don't like the look of but sound is something that happens to you whether you like it or not. Especially as someone with a visual impairment, sound becomes even more important. So if I feel like I belong in an environment, like I've been thought about in how that environment has been designed and I can be in it safely and comfortably then sound plays a very big part in that. It gives me the information I need to navigate and know where I am. Without it I am lost and disconnected.
>
> (Participant B)

Participant G discussed the notion of the auditory normate and described how as a neurodivergent person with hearing loss, changes in the sonic environment can significantly disrupt the equity of a space:

> I think that the design of most environments that I can think of, and that I am in regularly, doesn't give everybody equal opportunities. I think if you are someone who can move easily between loud and quiet environments without needing lots of prep or if you are someone who doesn't mind a cacophony of uncontrolled sounds like in a train station, if you are someone who can deal with these things then urban design is for you. Urban design expects you to be there. But if you're someone like me where sudden changes from loud to quiet or layers of different types of sound are incredibly overstimulating then I don't think urban design gives you an equal opportunity.
>
> (Participant G)

Participant D highlighted the importance of multi-sensory approaches to the accessible communication of information in the built environment from their experiences as a deaf person:

> The places where I feel I belong the most are those where I don't rely on sound to get information. So for example, if key things that I need to know are presented visually via text or images that works for me, but things like train announcements which are just sound are not accessible. If I see everyone leave a platform, I can guess that they have had some information via sound that I haven't had. So any sound-only information or interaction isn't accessible or equitable for me but things that are multisensory are.
>
> (Participant D)

Participant A also detailed how sound to them is information and impacts their temporal understanding:

> I use sound as my visual replacement. Whether that is the sound of rain on leaves, that tells me that it's spring, and the leaves are out. Or the sound of a dustbin lorry at the end of the road, telling me that it's Monday morning. Those things aren't noise, they are information for me that allows me to position myself in time. But when they're all blocked out by something like an aeroplane going over, it stops you knowing what point in the day you're in, or what's going on around you. It removes all sorts of pieces of information that allow me to orientate myself. If I am standing talking in a street and a very loud lorry goes by, I can't hear the person's voice anymore so they basically disappear, they suddenly vanish! So sound in those scenarios is the thing that most impacts my sense of belonging to a place.
>
> (Participant A)

Jordan Lacey (2016, 175) writes that in

> our cities, where everyday life has been reduced to the fulfilment of certain functions, the soundscape has come to reflect equivalently reductive environmental conditions. The whirr of climate controllers and hum of traffic, the yammering of construction sites and screeching of sirens are the expressions of a specific set of functional parameters to which social life must succumb.

The lived experiences of sound and soundscape design shared by research participants show how the urban soundscape that Lacey presents is not only reductive but also brings an oppressive power. A power to disrupt feelings of belonging and equity, prioritising citizens who conform to the normative expectations of the urban sound world and excluding those who don't.

Theme 2: Sound and Energy

There is a growing interest within the field of critical disability studies and inclusive design for research and practice which supports horizontal access and the design of environments where inhabitants can manage fluctuating energy levels. Horizontal access provisions are those which support people to rest within the built environment in ways that work for their body and mind. As disabled artist Raquel Meseguer notes the "horizontal body is not welcome in public … [it is] seen at worst as a threat, and at best as an inconvenience" (Meseguer 2022, 366). Designing horizontal access provisions is therefore a radical attempt to challenge the normative expectations of "how to be" in public space and an opportunity to re-frame resting, lying down, stretching out and being comfortable as a core tenant of inclusive practice. Both participants F and I discussed the impact that sound can have on their energy levels and how this influences their feelings of equity. Participant F described managing sound and energy as a neurodivergent person with chronic fatigue, whilst soundwalking through a quiet street near the river in central Glasgow, she noted:

> The biggest thing [when thinking about equity] is my energy. Has the space been designed to support me to manage my energy carefully? If a place like a big exhibition has lots of soft seats, then I know I can sit and rest and stay in that space for a long time. But if it's too loud then I just can't stay in there for very long. It makes me really need to focus on what's happening and my energy is completely zapped. I often see people sitting in noisy places and I think - gosh you can stay here as long as you want because this place isn't draining your energy but I feel like I'm being fried every second by the sounds bouncing off the walls.
>
> (Participant F)

Participant I spoke about the impact that sound can have on their energy levels as a neurodivergent person who is also a wheelchair user. Whilst soundwalking through a busy visual art gallery in Sheffield, she discussed the messages that the built environment sends to disabled visitors through different sensory modes:

> So we're in this kind of white foyer space and it's quite nice because there's lots of room to manoeuvre. I can look down there and I know I can fit through that space in my chair. I can see a lift so I can get upstairs and I've spotted the accessible toilet. So this place feels like somewhere I belong kind of visually. I'm getting lots of messages about me being welcome here from what I can see. But I'm not sure the sound of this place is giving me the same reassurance that this is going to be a space that works for me. I can hear a lot of things right now; people talking, the hand dryers blaring out the toilets, people in the cafe and the coffee machine – it's all just a messy blob of sound that is quite confusing. We're trying to have this conversation and I'm really focusing on your lips moving and trying to listen to the words that you're saying but I'm really having to concentrate. So I think the message I'm getting from the sound here is that this is going to be quite a tiring place to be in and probably I couldn't stay here for very long.
>
> (Participant I)

These narratives highlight how disabled people continue to experience barriers to the built environment because of the impact that sound and soundscape design have on their energy levels. Considering how to design sonic spaces that are manageable and support differing energy levels therefore creates an opportunity to disrupt the politics of visibility and design more equitable urban environments.

Theme 3: Sonic Information

Hamraie (2013) contends that the built environment is non-accessible to disabled people not only through physical and structural elements (e.g. stairs) but also through a lack of information relating to sensory elements (e.g. bright lights, intense smells or confusing layouts). This perspective has given rise to a growing number of accessibility provisions designed to provide clear information about the sensory components of an environment in advance of visiting. For example, many cultural institutions create a "Visual Story" – a document which uses images and text to show what a building looks like. Research participants discussed the importance of knowing about sound in advance of visiting a location. Participant L said:

> I love it when you can see like on Google how busy places are because I tend to prefer to be in noisier places. I can find quiet quite unnerving. It can make me twitch and kind of agitated because I sort of feel like the place is telling me what to do. Like the room is saying "Shush! Be still! Don't make noise". So knowing before I go somewhere if it's going to be busy or loud is one thing that makes it feel equal and fair to me.
>
> (Participant L)

Participant G shared some techniques she uses for learning about the sounds of an environment in advance:

> When somewhere feels unequal it's often because I don't know what it's going to sound like. I watch YouTube videos of things like train stations or people walking down a particular street to get a sense of the sound before I go. Watching videos gives me a chance to hear what a place might sound like so I know more about some of the sounds I might come into contact with. I can get very dizzy and off balance because of loud sudden noises like a drill on a busy road but I don't mind being in some noisy places if I can prepare properly.
>
> (Participant G)

Participant H described the strategies for visiting new spaces that they have developed as an autistic person who experiences hyperacusis and misophonia:

> Going to a place for the first time can be triggering for me, I don't know what it will sound like. Certain sounds might be too much for me, the uncertainty makes me anxious. Often, I will get someone to go into a place first and report back or make recordings on their phone and message them to me. But nowhere really provides this information, it's always something I have to find out for myself.
>
> (Participant H)

Building on the aforementioned format of the Visual Story, a new accessibility provision titled a "Sonic Story" has been developed. A Sonic Story is an infographic that shows how the sound levels of an environment or event change over time (Renel, 2019). The technique was created by the disabled-led company Touretteshero through doctoral research led by the author at the Helen Hamlyn Centre for Design. Touretteshero regularly designs and delivers Sonic Stories for cultural institutions across the UK including the National Theatre, The Royal Shakespeare Company and Shakespeare's Globe Theatre. The Sonic Story provision is an exemplar of how increasing the availability of information about the sounds of a built environment can improve the equity of the space for disabled and neurodivergent people.

Theme 4: Surfaces, Height and Light

The design and management of the built environment serve as a functional and communicative vessel for the ideology and norms of the people who made it (Renel, 2019). Design frames who can do what and on what terms within the politics of space and place (Certoma et al., 2012). Within inclusive design, there is an acknowledgement that considerations within design processes are not value-neutral but represent the realities and possibilities of how a space will be shaped socially. Hamraie (2017, 19) writes that from "a doorframe's negative space to the height of shelves and cabinets, inhabitants bodies are simultaneously imagined, hidden and produced by the design of built worlds." Research participants expanded on this idea by sharing lived experiences relating to how the physical and visual environment (e.g. the height of objects, the materiality of surfaces and light levels) can impact their sonic experiences and feelings of equity. Participant I commented:

> The height of things that emit sounds like intercoms or speakers is set by building standards or regulation, but there is a clear assumption about 'the right height' for listening to these things. For loads of disabled people who are little, in a wheelchair or for whatever reason don't match up to these assumptions these things aren't accessible.
>
> (Participant I)

Participant J discussed their experiences of sound and height whilst soundwalking down a busy street in central Sheffield:

> I love this High Street it's definitely a place where I feel like I belong. It's near my house, it's full of life, people, traffic and the sound makes me feel alive. [As a wheelchair user] I'm often lower than most people, I exist more on the level of buggies and dogs then I do on the level of lots of adults so I find that my sound world is quite low compared to other people's. And designed stuff doesn't always match up to my world. There are obvious things like if I need to speak to someone to use an information

> point or to order a drink but also, I think the built environment sends messages more generally with things like the height of light switches and buttons and intercoms. That message is that a certain person has been thought about and so others like me who don't match up are less important.
>
> (Participant J)

Participant C discussed the role of surfaces which reflect sound in different ways from their perspective as a neurodivergent person who experiences voice hearing:

> I am pretty sensitive to sound, particularly high-pitched noises. So if I'm somewhere where there are high pitched sounds and noises or where sound is bouncing off surfaces like metal where it gets high pitched and tinny, I find that very overwhelming. Reflected sound can also trigger my voices so a busy external environment can make my internal environment feel really busy too.
>
> (Participant C)

Participant D talked about the interplay between different sensory modes in shaping their feelings of equity in the built environment:

> Sometimes, as a Deaf person accessibility can be thought of only to do with hearing or what is audible. But actually for me it's about all the senses. If the lighting is good I can lip read but if the lighting is bad and I can't see people's faces then conversations and interactions become impossible. So really one of the things that has the biggest impact on me feeling like sound is accessible and the environment is equal is the lighting and how this affects communication.
>
> (Participant D)

Participant A described his local neighbourhood in North-East London and the impact the physical environment has on his sonic experiences as someone with a visual impairment:

> When I'm trying to cross the road at the top of our road, at the junction, there's a raised table (where there used to be a curb), so I can't always be sure whether I'm on the road or on the pavement, I just have to guess it. Sound there combines with the loss of pavement information and I'm not sure where I'm standing. Add in the new phenomenon of very quiet electric scooters, very quiet electric cars and electric motorbikes. This combination has caused every little fleeting sound that could be wheels moving quickly over tarmac to become an immediate threat. When actually 9 times out of 10 it's not. For me, electric propulsion has removed my ability to identify and locate moving objects around me and that's had a

> hugely negative impact on whether I feel the city where I live is an equitable place.
>
> (Participant A)

In writing on neurodivergence and architecture, Clarke et al. (2022, xviii) suggest that architecture and the built environment "are ethically relevant because they enact a generalized set of norms about how humans should conduct themselves, and in doing so, they constrain, enable, and transmute the actions of individuals and collectives in ways that significantly impact their well-being". The lived experiences shared by the participants bring a sonic consciousness to such considerations of the ethical relevance of the built environment – challenging the everyday design practices that perpetuate patterns of inequality towards more sonically inclusive design.

Theme 5: Diverse Sonic Spaces

Blesser and Salter (2019) describe the difference between an acoustic architect (whose focus is the way that the space changes the physical properties of sound waves) and an aural architect (whose focus is the way that listeners experience the space). Research participants described how the built environment can become more equitable by having spaces with different sound levels and acoustic profiles easily available and clearly signposted. This highlights the importance of bringing together both acoustic and aural architects in the creation of equitable sonic spaces. Participant H described how they have developed their own individualised maps of London which enable them to navigate the city as an autistic person who experiences hyperacusis and misophonia:

> When I get to know parts of the city well, I know where the quiet and loud bits are and that allows me be there confidently. Like there is a great book shop near Oxford Circus station which is always quiet so if I'm getting too stimulated by Oxford Street then I know I can go there.
>
> (Participant H)

Whilst soundwalking through a busy gallery in Sheffield, participant J discussed the significance of where information is positioned within the built environment:

> So we are in a room now which is so much nicer [than the foyer] it's quiet. As we came through, we were talking about the difference in visual and hearing experiences and there was a map by the door that was really beautiful and big. I think it even had tactile parts so you could understand it using touch. The map had visual information like colour coding for each floor of the gallery but nothing about sound like where is quiet or loud. Often visual information is available but nothing to do with sound. Also, the map itself was in a really loud room so I was struggling

> to look and understand it because it was so noisy. If it was in here [where it's quieter] I could sit and look at that map for ages and plan my visit.
>
> (Participant J)

Participant E talked openly about the power that individual sounds can have in disrupting their experience of the built environment as a neurodivergent person with Meniere's disease:

> One of the things that [Meniere's] does is if I hear certain sounds or particular frequencies, like often high frequencies, it can trigger some very debilitating symptoms like vomiting, severe sweating and dizziness. So, I can be trying to do something that's meant to be fun like going shopping with a friend or going to see something in a gallery or whatever and a sound (like an announcement or sudden beeping) can just change that from being something nice to something horrible. And if the sound makes me react my first question is always "where do I go"? The only place that I regularly find to run away to is the toilets.
>
> (Participant E)

The narratives shared by participants signal the importance of bringing together the work of both acoustic and aural architects – to ensure a diversity of sonic spaces are available and communicated in an accessible manner.

Conclusion

This chapter has described how the concepts of "aural diversity" and the "auditory normate" have so far been underutilised in discourse and discussion surrounding spatial inequality and equity in the built environment. The counter-narratives shared by participants across all four themes above highlight a unique set of auditory experiences that disabled and neurodivergent people continue to navigate because of the current design and management of the built environment. It is clear that the experiences of disabled and neurodivergent people have a pivotal role to play in challenging normative assumptions about human experience in the built environment and creating new opportunities for more equitable environments to emerge. Recent policies in relation to accessibility (BS8300-2 2018), acoustics (ISO 12913-1, 2014; Kang et al., 2013) and urban soundscape design (Kang and Schulte-Fortkamp, 2016) all signal positive changes with regard to the design and management of more inclusive urban sound and soundscapes. From a theoretical standpoint, LaBelle (2019) suggests the notion of acoustic justice as a set of practices that intervene within normative conditions by

> upsetting and reworking a given socio-materiality through more energetic, animate, and ephemeral forces and forms. Materiality, and the spatial or demographic contours of the built environment, are detuned and

retuned, shaken and rattled through a range of expressions and appropriations that lead to a reworking of the heard as a project of acoustic justice.

(LaBelle, 2019, 178)

Increasing the understanding of aural diversity and the auditory normate within the policies and practices manifested in the built environment and mediated spatially therefore creates an opportunity to foreground and celebrate the diverse auditory experiences of disabled and neurodivergent people in a move towards a more socially and sonically equitable future.

References

Auraldiversity Network (Website). 2023. Available: www.auraldiversity.org [Accessed 01.12.23].

Berger, R. J. and Lorenz, L. S. 2017. *Disability and Qualitative Inquiry: Methods for Rethinking an Ableist World*. New York: Routledge.

Blesser, B. and Salter, L. 2019. *Spaces Speak: Are You Listening? Experiencing Aural Architecture*. Cambridge, MA: MIT Press.

BS 8300-2. 2018. *Design of an Accessible and Inclusive Built Environment - Part 2*. London: British Standards Institution.

BS ISO 226. 2003. *Acoustics – Normal Equal- Loudness-Level Contours*. London: BSI.

Certoma, C., Clewer N., and Elsey D. (eds.). 2012. *The Politics of Space and Place*. Newcastle: Cambridge Scholars Publishing.

Clarke, A., Boys, J., Gardener, J. 2022. *Neurodivergence and Architecture*. Amsterdam: Elsevier Academic Press.

Daza, S., and Gershon, W. S. 2015. Beyond Ocular Inquiry: Sound, Silence, and Sonification. Qualitative Inquiry, Vol 21(7).

Delgado, R. 1989. Storytelling for oppositionists and others: A plea for narrative. *Michigan Law Review*, 87(8), 2411–2441.

Dovey, K. 1999. *Framing Places: Mediating Power in Built Form*. New York: Routledge.

Drever, J., and A. Hugill. 2023. *Auraldiversity*. New York: Routledge.

Drever, J. L. 2017. The Case for Auraldiversity in Acoustic Regulations and Practice: The Hand Dryer Noise Story. Presented at the *24th International Congress on Sound and Vibration*, London.

Ellis, K., Garland-Thomson, R., Kent, M., Robertson, R. 2019. *Manifestos for the Future of Critical Disability Studies*. New York: Routledge.

Hamraie, A. 2013. Designing collective access: A feminist disability theory of universal design. *Disability Studies Quarterly*, 33(4). Available: http://dsq-sds.org/article/view/3871/3411 [Accessed 01.12.23].

Hamraie, A. 2017. *Building Access: Universal Design and the Politics of Disability*. Minneapolis, MN: University of Minnesota Press.

ISO 12913-1. 2014. *Acoustics – Soundscape – Part 1: Definition and Conceptual Framework*. Geneva: The International Organization for Standardization.

Kang, J., Chourmouziadou, K., and Sakantamis, B. 2013. *COST Action TD0804 – Soundscape of European Cities and Landscapes*. Brussels: The European Cooperation in Science and Technology.

Kang, J., and Schulte-Fortkamp, B. 2016. *Soundscape and the Built Environment*. Boca Raton, FL: CRC Press.

Kanngieser, A. 2011. A sonic geography of voice: Towards an affective politics. *Progress in Human Geography*, 36(3): 336–353.

Labelle, B. 2019. *Acoustic Territories Sound Culture and Everyday Life*, 2nd edition. London: Bloomsbury.
Lacey, J. 2016. Sonic Rapture: A Practice-led Approach to Urban Soundscape Design. London: Bloomsbury.
Meseguer, R. Z. 2022. Chronic pain & chronic illness: A crash course in cloudspotting. In: Clarke, A., Boys, J., Gardener, J. *Neurodivergence and Architecture*. Available here: https://www.sciencedirect.com/science/article/pii/S2589295922000054
Neuman, W. 2011. *Social Research Methods: Qualitative and Quantitative* Methods. Boston: Allyn and Bacon.
Raines, A. R., Francis, G. L., Fujita, M., & Macedonia, A. 2023. Belonging from the perspectives of individuals with disabilities: A scoping review. *Psychology in the Schools*, 60, 2112–2127. https://doi.org/10.1002/pits.22864
Ravelli and McMurtie. 2016. *Multimodality in the Built Environment: Spatial Discourse Analysis*. New York: Routledge.
Reed, S., Bohr, J. 2021. The influence of local built environments on social wellbeing: a community's experience with social isolation, social loneliness, and social belonging. *International Journal of Community Well-Being* 4, 393–413. https://doi.org/10.1007/s42413-020-00091-6
Renel, W. 2018. Auraldiversity: Defining a hearing-centred approach to socially equitable design of the built environment. *Built Environment*, Vol 44(1): 36–51.
Renel, W. 2019. Sonic accessibility: Increasing social equity through the inclusive design of sound in museums and heritage sites. *Curator: The Museum Journal*, 62(3): 377–402.
Renel, W. 2023. *The Auditory Normate: Engaging Critically with Sound, Social Inclusion and Design*. In Drever, J., and Hugill, A. Auraldiversity. New York: Routledge.
Rieger, J. 2023. Design, Disability and Embodiment Spatial Justice and Perspectives of Power. New York: Routledge.
Sitter, K. C., & Grittner, A. L. 2021. When participatory approaches are inaccessible: A movement toward research engagement through multi-sensory storytelling. In *Centering Diverse Body Minds in Critical Qualitative Inquiry* (pp. 37–48). New York: Routledge.
Soja, E. W. 2011. Postmodern Geographies; The Reassertion of Space in Critical Social Theory. New York: Verso.
Solórzano, D. G., & Yosso, T. J. 2002. Critical race methodology: Counter-storytelling as an analytical framework for education research. *Qualitative Inquiry*, 8(1), 23–44.
Sterne, J. (2012). The Sound Studies Reader. New York: Routledge.
Stoever, J. L. 2016. *The Sonic Color Line: Race and the Cultural Politics of Listening*. New York: NYUP.
Westerkamp, H. 1974. Soundwalking. *Sound Heritage*, 3(4). Victoria B.C.
Yuval-Davis, N. 2006. Belonging and the politics of belonging. *Patterns of Prejudice*, 40(3): 197–214, https://doi.org/10.1080/00313220600769331
Zouras, E. 2020. *Belonging' as a Concept in Placemaking: Exploring Perceptions in Ikaria, Greece*. KTH Royal Institute of Technology. Available: https://www.diva-portal.org/smash/get/diva2:1592906/FULLTEXT01.pdf [Accessed 01.12.23].

14 Digital Equity in the United States

Disparities and Socioeconomic Influences on Internet Access and Use

Avijit Sarkar and James B. Pick

Introduction

Member States of the United Nations (UN) have committed to achieving sustainable development for all nations and populations from all segments of societies by adopting the 2030 UN Agenda for Sustainable Development. Reducing inequalities within and among countries is a key UN Sustainable Development Goal (UN SDG) and bridging the digital divide and advancing digital inclusion and equity are key priorities to promote social development and well-being for all. The adoption and diffusion of information and communication technologies (ICTs), particularly the internet, has continued to grow (Figure 14.1).

Similarly, there has been growth in the accessibility and use of other ICTs as measured by expansion in the base as well as the penetration rates of mobile-cellular telephone subscribers, fixed broadband subscribers, mobile broadband subscribers, population covered by at least a 3G mobile network, and fixed and mobile broadband traffic (ITU, 2024).

However, demographics and place of residence continue to remain important determinants of disparities in ICT access and use worldwide, in developed as well as developing nations. For example, in 2023 only 50.4% of rural populations worldwide used the internet, compared to 81.2% of their urban counterparts. This urban–rural gap in internet use is more pronounced for low and low-middle-income countries compared to upper-middle and high-income countries (ITU, 2024). Internet use disparities are also attributed to demographic factors such as age structure, gender, race-ethnicity, and educational attainment, to name a few. Within nations, developed and developing, these demographic factors along with economic, occupational, and infrastructural factors, the policy and regulatory environment, and societal openness have been documented to be key determinants of digital disparities, often at the state, provincial, and prefectural levels (Pick and Sarkar, 2015).

This chapter focuses on the evolution of the digital divide in the United States and discusses different levels of the U.S. digital divide through sociocultural and socioeconomic lenses, along with their implications for digital equity and inclusion.

DOI: 10.4324/9781032631110-15

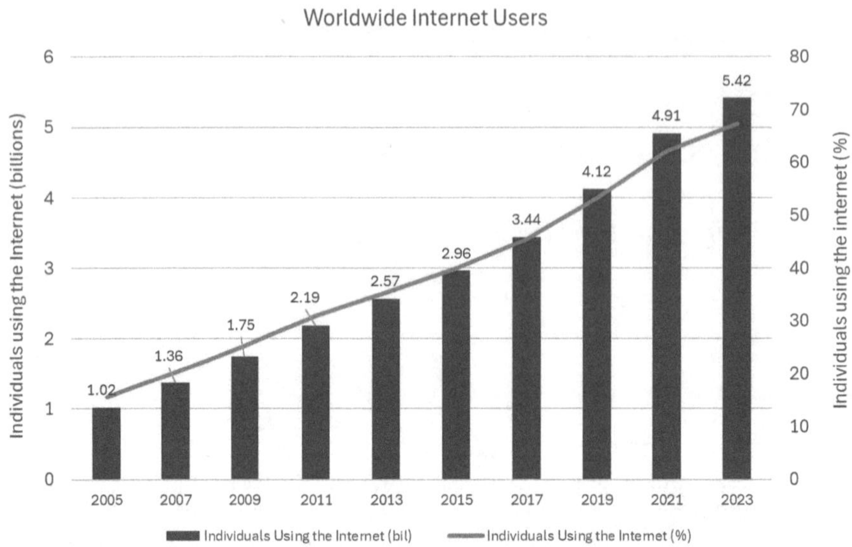

Figure 14.1 Worldwide internet use and penetration (2005–2023). Data Source: International Telecommunication Union (ITU). Credit: Avijit Sarkar.

Overview of the Digital Divide

The term "digital divide" refers to the "gap between individuals, households, businesses and geographic areas at different socioeconomic levels with regard both to their opportunities to access information and communication technologies (ICTs) and to their use of the Internet for a wide variety of activities" (OECD, 2001, p. 5). As evident from this well-accepted definition, the digital divide exists and has been studied by researchers and scholars at many levels. Digital divides are intrinsically tied to socioeconomic factors and socioeconomic as well as societal outcomes. In fact, the actual state of digital disparities is quite complex (van Dijk and Hacker, 2003) and is linked to social, economic, and cultural differences within and between societies and populations. As a result, researchers across several disciplines including sociology, education, political science, information systems, telecommunications, geography, law, and communication science have studied the intersectionality of variables such as gender, age, ability, race, tribe, caste, culture, geography, migration or displacement status, ethnicity, religion, education, language, socioeconomic status, freedom and liberties, and policy and regulations, among others, when studying and analyzing digital divides (Lythreatis, Singh, and El-Kassar, 2022; Scheerder, van Deursen, and van Dijk, 2017; van Dijk, 2020).

The digital divide is important since people cannot play a role in contemporary society without using digital technology including ICTs such as the internet. To gain access to contemporary education, healthcare, news and entertainment, commerce, and governmental services, using the internet is critical. The internet also enables access to a variety of online networks including social networks, professional networks, networks of household devices connected to the internet, as well as to contemporary economies including collaborative consumption-based sharing economies via which ridesharing services and short-term vacation rentals become available to individuals. Hence, it has been argued that digital divides can lead to social exclusion (van Dijk, 2020), civic disengagement (Qureshi, 2012), and arguably compromise the security of individuals (van Dijk, 2020), while stymieing economic growth, development, and competitiveness of nations (van Dijk, 2020).

The digital divide has been conceptualized at three levels (Lythreatis, Singh, and El-Kassar, 2022; Scheerder, van Deursen, and van Dijk, 2017). The first level pertains to internet and general ICT access while the second level focuses on digital skills and use. The second level of the digital divide (Hargittai, 2002) was introduced based on arguments by scholars and policymakers that having material access to devices such as computers, laptops, and mobile phones as well as the internet was not meaningful if individuals lacked digital literacy, skills, knowledge, support, and training. The third level of the digital divide is tied to tangible outcomes centered around effects and impacts on people, organizations, and society at large. Outcomes have been measured in economic, social, political, cultural, educational, and personal development domains (van Deursen and van Dijk, 2014), as well as for individual well-being (Helsper, van Deursen, and Eynon, 2015). Each of these levels have been studied using a variety of theories grounded in disparate disciplines at the levels of individuals, households, businesses, and geographic areas, including but not limited to nations, sub-national geographies such as states, provinces, prefectures, districts, municipalities, and cities (Grubesic and Helderop, 2022; Pick and Sarkar, 2015; Rahman, Ara, and Chakma, 2022; Setthasuravich and Kato, 2022; Song, Wang, and Bergmann, 2020). Next, this chapter provides an overview of the digital divide in the United States, with a focus on access and use of the internet.

Digital Divide in the United States: Internet Access

In 2023, the United States ranked seventh among 170 nations worldwide in terms of ICT development, as measured by the ICT Development Index (IDI) of the International Telecommunications Union. The United States was preceded by smaller nations Kuwait, Singapore, Qatar, Denmark, Estonia, and Finland, of which Denmark's 2022 estimated population (approx. 5.9 million) was the largest. In November 2023, the National Telecommunications and Information Administration (NTIA) of the U.S. Department of Commerce estimated 266.47 million internet users in the United States, which is 83.3% of all civilians aged 3 years and above (NTIA, 2024). Compared to 2021, almost 13 million or 3% more people (civilians aged 3+) reported using the internet in the United States in 2023. This post-pandemic growth in the base of internet users marks a period of progress that rivals the pre-pandemic 2015–2017 period which witnessed similar growth. Encouragingly, much of the growth in the internet-connected population came from demographic groups that had traditionally been considered to be on the wrong side of the U.S. digital divide. The NTIA has showcased American Indians and Alaska Natives, long considered laggards in internet and digital connectivity, to be one example of a demographic group whose internet adoption increased from 75% in 2021 to 83% in 2023, matching the national penetration rate. Despite this progress, almost 12% of people lived without any internet connectivity in 2023 (Goldberg, 2024). Major reasons for non-adoption of internet in the United States range from lack of motivation or need to use the internet, concern about affordability, lack of adequate computing in the household (computer, laptop, etc.), and privacy and security concerns (NTIA, 2024).

In fact, race/ethnicity is one of several demographic dimensions to which digital disparities have been attributed to. Other demographic dimensions include age structure, educational attainment, sex, population density, and presence of children in the household. For example, 90% of all U.S. internet users aged 15+ hold a college degree versus 81% of those who are high school graduates. Economic factors attributed to disparities in internet access include employment status, family income, and type of occupation (NTIA, 2024). Keeping in mind the economic impact of high-speed internet, particularly broadband access on local economies, including any effect on small businesses or jobs, the Access Broadband Act of 2021 was operationalized as a dashboard by the NTIA and the U.S. Census Bureau. The dashboard enables visualization of interactions between broadband access and social and economic indicators including statistics on employment, small business establishments, wages and income, poverty, home values, population change and migration, educational attainment, and Gross Domestic Product (GDP).

Using the Access Broadband Dashboard, for example, we can see that the proportion of households in U.S. counties reporting at least one type of internet subscription (i.e., broadband such as cable, fiber optic, or DSL; a cellular data plan; satellite; a fixed wireless subscription; or other non-dial-up subscription types) is positively associated with year-over-year change in real GDP. For instance, Los Angeles County, California, with an almost 14% higher proportion of the population with access to 25/3 Mbps broadband (as of December 31, 2019) than Imperial County, California, reported a 7.9% increase in real GDP from 2020 to 2021 (Access Broadband Dashboard, 2023) (Table 14.1). Imperial County, on the other hand, reported a 0.4% decline in its real GDP over the same period. The same is true for other economic indicators such as labor force participation, establishment entry rate, and median home value.

Analyzing interactions and statistical associations between broadband availability and such economic indicators can be the starting point for empirical studies that investigate the effects of broadband access disparities (first level of the digital divide) on outcomes (third level of the digital divide). As discussed earlier in this chapter, digital divides are first and foremost technological, yet digital disparities must be studied through the lens of the

Table 14.1 Broadband availability in Los Angeles and Imperial Counties, CA, with sociodemographic and economic attributes.

Broadband availability and sociodemographic and economic attribute (various years, 2017–2021)	*Los Angeles County, CA*	*Imperial County, CA*
Households with a broadband subscription	88.9%	85.3%
Population with access to broadband services of at least 25/3 Mbps	99.5%	85.7%
Urban population (2020 U.S. census)	99.11%	81.59%
Employed	60.3%	43.7%
Labor force participation	64.8%	50.8%
Unemployed	8.9%	17.3%
Annual change in employment	3.5%	5.1%
Workers self-employed	8.9%	6.6%
Workers that work from home	10.9%	6.1%
Weekly wage	$1,488	$915
Median household income	$77,356	$50,970
Poverty (American community survey)	13.9%	21.0%
Establishment entry rate	12.6%	8.0%
Annual change in establishments with less than 20 employees	2.0%	−1.3%
Annual change in establishments with less than 500 employees	1.7%	−1.9%
Annual change in real gross domestic product (GDP)	7.9%	−0.4%
Median home value	$647,000	$219,800
Annual change in population	−1.6%	0.2%
Net migration rate (per 1,000 population)	−17.7	−2.2
Population with a bachelor's degree or higher	34.0%	15.2%
High school-aged population not enrolled, not a graduate	3.0%	2.0%

Data Source: Access Broadband Dashboard, 2023.

intersectionality of social, economic, cultural, and political forces (Selwyn, 2004). This can help policymakers and the public to assess how changes in broadband availability and adoption, as a standalone factor, or alongside human capital, labor, multifactor productivity, and other economic factors, could influence local economies including the resilience to disruptions such as the shutdowns during the COVID-19 pandemic.

Over the years, a handful of studies have shed light on important causal relationships between the U.S. digital divide and a host of underlying factors. A recent county-level study of the U.S. digital divide (Sarkar, Pick, and Rosales, 2023) found that access and use of ICTs (computer ownership in the household, high-speed internet, broadband, and mobile internet subscription in the household) vary significantly in metropolitan, micropolitan, and rural counties in the United States reinforcing that the urban–rural digital divide has continued to worsen (White House, 2015) despite the progress achieved during the COVID-19 pandemic. Age structure, college education, working-age population, and race-ethnicity were found to be determinants of ICT access and use by county. Previously, Khatiwada and Pigg (2010) examined spatial patterns of internet diffusion in U.S. counties and found that market demand is the driving force for internet adoption in urban counties due to a more educated population, the presence of business establishments, and higher housing values. The role of social capital in bridging the digital divide in the United States has also been studied (Chen, 2013; Pick and Sarkar, 2015). It has been found that different forms of social capital facilitate and provide material and skills transfer from the digital natives to those who are digitally less skilled and perhaps do not own their own devices and technology. Apart from these studies, the NTIA's Digital Nation series of reports (NTIA, 2013, 2014) have documented sociodemographic and socioeconomic dimensions of the access divide in the United States. The findings of these studies and reports reinforce the intersectionality of digital disparities with social, economic, and cultural differences in the United States.

It is important to note that while most Americans have access to some form of broadband internet connection, the bandwidth, the quality of service, the number of broadband providers, and the availability of choice among households remain highly uneven throughout the country. Across the country, urban dwellers often have the option to choose between various broadband technologies such as digital subscriber line (DSL), fixed wireless, cable, and fiber optics at various speed tiers from multiple providers, but residents of rural areas are hamstrung by limited platform and provider choices (Grubesic and Helderop, 2022). For example, in 2022, 44.6% of the U.S. urban population had access to ultrafast 940/500 Mbps fiber-based high-speed broadband compared to the 23.9% of rural population (40.4% nationwide). In contrast, 98% of the urban population had access to broadband speed of 100/20 Mbps in 2022 compared to 72% of the rural population (FCC, 2024). However, despite the availability of faster speeds resulting from fiber-optic availability in urban areas, affordability remains an issue, causing urban broadband subscribers to choose lower-speed tiers. In fact, some of the same sociodemographic factors associated with low internet adoption rates such as age, urbanization, race/ethnicity, and income cause disparities in actual internet speed (Gallardo and Whitacre, 2024). These disparities have implications for purposeful use of the internet (e.g., data-intensive activities such as online learning, teleworking, and telehealth) as well as the allocation of funding for broadband infrastructure expansion, and the consequent advancement of digital inclusion and equity, which is discussed later in the chapter.

Globally, the adoption and use of the internet and other information and communications technologies have been studied extensively by multinational organizations such as the International Telecommunication Union (ITU) as well as by national statistical and telecommunications agencies. The ITU's well-known IDI serves as a tool for policymakers to assess the state of digital development by measuring the connectivity of people, households, communities, and businesses and the enablers of connectivity including infrastructure, affordability, devices, skills, and safety and security. Studies have shown that many of the demographic and socioeconomic factors influencing internet access disparities in the United States also exert varying levels of influence on internet access and use in China, India, Japan, EU nations, and African countries, all of which have varying levels of digital development (Grishchenko, 2022; Pick, Ren, and Sarkar, 2024; Pick and Sarkar, 2015). In regions and nations where internet censorship has been used as a tool for political repression, suppression of civil rights, prevention of publication of corruption, or deterrence against the spread of rumors and misinformation that may lead to social unrest (often at the behest of the state; Adams and Warf, 2022), it has been found that the extent of societal openness is often correlated with varying levels of internet access and use (Pick and Sarkar, 2015; Warf, 2013). Other region or nation-specific factors that have been found to be associated with internet disparities include the proportion of English-speaking population (in Latin America), infrastructural malaise including unreliable electric grid, extent of human development, and effectiveness of ICT policies and regulations (in sub-Saharan Africa) (Pick, Sarkar, and Parrish, 2021; Pick and Sarkar, 2015). In short, the underlying factors contributing to global digital divides are often demographic and socioeconomic, but region and country-specific differences have implications for the formulation of policies that, through changes to the built environment and other means, can advance digital equity and inclusion.

Next, we turn to the second level of the U.S. digital divide, which pertains to the use of internet for various purposes, their disparities, and underlying factors.

Digital Divide in the United States: Internet Use

Since the mid-2010s, the NTIA has been systematically tracking the spectrum of online activities of American internet users. Sixteen different indicators of online activities spanning social networking, online banking and bill payment, e-commerce, e-entertainment, online education and training, and accessing sharing economy platforms are some of the activities that have been tracked by the NTIA. As depicted in Figure 14.2, there has been growth in internet use across a broad spectrum of online activities. For example, between 2015 and 2023, the proportion of American internet users aged 15+ using online financial services grew from 64.60% to 76.10%, while the growth in participation in online video and voice calls has almost doubled, from 36.90% in 2015 to 66.10% in 2023. This dramatic increase was likely fueled by the rapid proliferation and increase in the use of online video conferencing platforms such as Webex, Zoom, and Microsoft Teams during the COVID-19 pandemic.

However, there is considerable variation in the proportion of American internet users who have participated in such online activities. For example, in

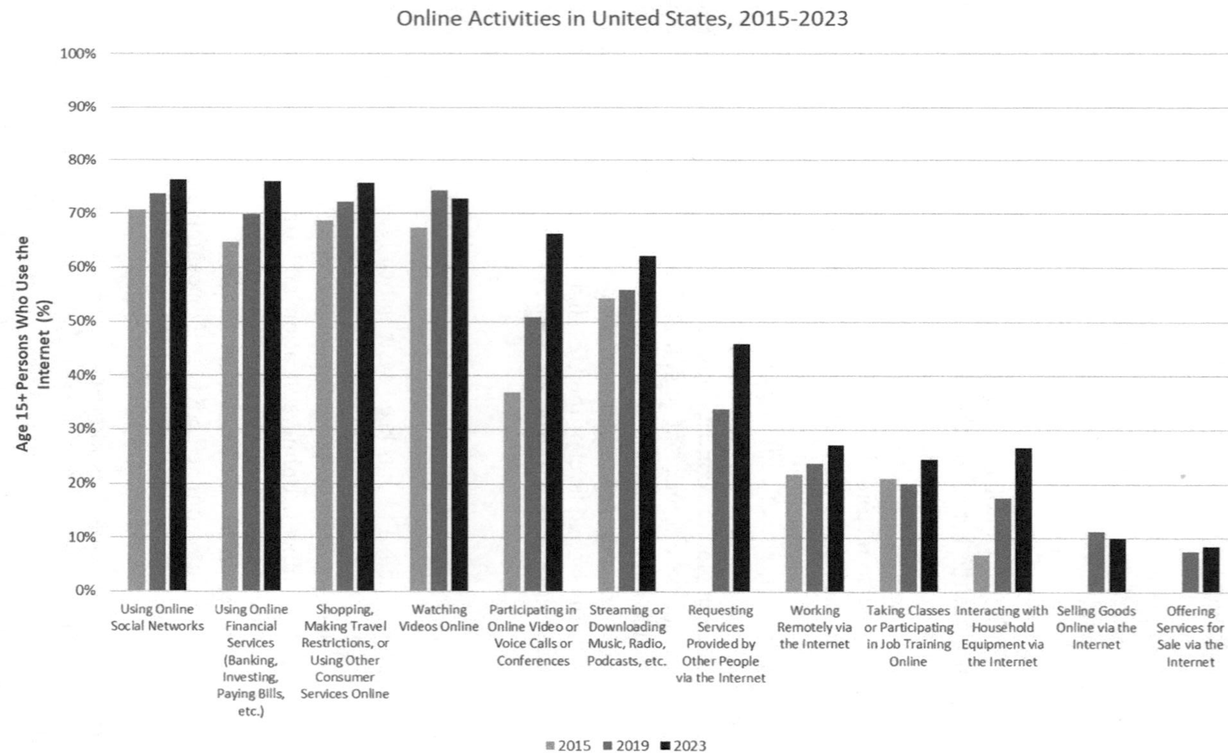

Figure 14.2 Longitudinal changes in online activities of American internet users, 2015–2023. Data Source: National Telecommunications and Information Administration. Credit: Avijit Sarkar.

2023 more than three-quarters of American internet users shopped, made travel reservations, and used other consumer services online, in contrast to a little over one-quarter who worked remotely using the internet (NTIA, 2024). Such gaps are even more pronounced for activities that are likely to require higher levels of online skills such as using the internet to connect to household devices including internet-connected thermostats, utility meters, lights, plug-ins, or other internet-connected solutions such as virtual assistants in the form of a smart speaker.

Significant geographic disparities exist within the United States across the spectrum of online activities. For example, to access online financial services (online banking, investing, and bill payment), north-eastern states such as New York, New Jersey, Pennsylvania, and Maryland along with Nevada in the West reported the lowest usage (by no more than 55% of the state population aged 15+) in November 2021 (Sarkar and Pick, 2023). 56–60% of the population in populous states such as California, Texas, Florida, Michigan, Virginia, West Virginia, and neighboring northeastern states such as Connecticut, Massachusetts, and New Hampshire engaged in this online activity during the same period. Moderate levels of engagement (61–65% of the state population aged 15+) were most prevalent across U.S. states and were observed for states in the Pacific Northwest, Rocky Mountains, Southwestern United States, and extended parts of the Deep South stretching from Louisiana in the Gulf Coast to South Carolina in the Atlantic coast. Moderate levels were also noticed for parts of the Midwest (Indiana, Ohio) and Appalachia (Kentucky). High levels of use of online financial services (66–70% of the population aged 15+) was observed in a couple of southern states (Arkansas and Tennessee), Midwest, and upper Midwest (from Illinois to Minnesota), while the highest use (70–87% of the state population) was reported for Iowa in the Great Plains, South Dakota in the Prairie region, and Washington, D.C (Sarkar and Pick, 2023).

While Washington, D.C.'s population is entirely urban, there may be a reversal in terms of the urban–rural divide that is often associated with broadband and internet use in the United States. In fact, the laggards (states such as New York, New Jersey, Massachusetts, Connecticut California, and Florida) have high to very high levels of urbanization, with 70% of the state population living in urban areas. Conversely, except for Washington D.C., the leading states have a much lower proportion of urban residents (Sarkar and Pick, 2023).

Disparities are also observed across demographic indicators such as age, sex, race/ethnicity, family income, education, and employment status. For example, for shopping, making travel reservations, and using other consumer services online, 79.3% of White and 77.6% of Asian internet users engaged in this online activity compared to 71% of African American and 65.8% of Hispanic internet users in 2023. There is a considerable gap between age-based cohorts, with 82.3% of internet users aged 25–44 engaging in this activity versus 64.7% of those aged 65+ in 2023 (NTIA, 2024). To request services provided by others online, the urban–rural gap is significant in 2023 at 48.8% for urban versus 26.1% for rural internet users. For streaming or downloading music, radio, and podcasts, educational attainment is a key differentiator, with 52.9% of internet users with high school diplomas engaging in this activity in 2023 compared to 70% of internet users who are college graduates (NTIA, 2024).

Similar to disparities between U.S. states, there are considerable differences among U.S. counties in the use of the internet for online activities. A recent study on social networking in U.S. counties found that Facebook use among American social media users was high (53–74% of adult users, last 30 days) in bicoastal, largely urban, metropolitan counties in California, Florida, large parts of the North-eastern seaboard, the Great Lakes region, as well as the Rocky Mountain states. Low levels of social networking via Facebook were more common in the rural South as well as the prairie states. While similar patterns were also observed for other social media and networking platforms such as LinkedIn, Twitter, YouTube, and Instagram, their intensity of use was significantly lower than Facebook (Sarkar, Pick, and Raut, 2021). In addition, it has been found that of the 152 rural counties that have been identified as statistically significant outliers in Facebook usage, several are home to universities and military bases. It has been reasoned that educational institutions such as universities, national labs, and military bases are likely to be home to frequent users of ICTs including social media. Hence, even though outlier counties are in rural areas, the presence of institutions of higher education and research catalyzes social media access overcoming their locational "disadvantage."

A recent study has shown that the proportion of remote workers is highly positively associated with the extent of online activities in U.S. states in 2021, while median age is inversely associated with a broad spectrum of online activities (Sarkar and Pick, 2023). It was reasoned that as more people worked from home during the pandemic, they increasingly relied upon the internet for e-communication, e-entertainment, teleworking, e-education, as well as e-commerce. Curiously, however, the proportion of the urban population was found to be inversely associated with using email, text messaging and instant messaging, watching videos online, and streaming or downloading music, radio, and podcasts. It is likely that the population migration that took place during the pandemic from densely populated, large urban metropolitan areas to less densely populated micropolitan or rural areas contributed to a reversal of the urban–rural divide for a selection of online activities. Whether this reversal endures remains to be seen. The urban–rural divide along with the influence of other factors has implications for digital equity, as discussed below.

Implications for Digital Equity and Inclusion

The digital divide is a complex and multidimensional phenomenon. Considering the disparities in access, use, and outcomes, and the underlying factors, it is essential to consider a geographic approach to the study of digital divides (Warf, 2021). This is particularly critical to study, understand, visualize, and critically analyze the intersections between digital, social, cultural, and economic equity. From a methodological perspective, to understand the factors that have causal relationships with ICT access and use, it is critical to first understand the role that geographic proximity (nearness) and relatedness may play. After all, as "birds of a feather flock together," it is quite likely that individuals, households, businesses, and geographic areas that have similar levels of digitalization to their neighbors also share similar demographic, cultural, social, and economic characteristics. Therefore, it is methodologically appropriate to measure the extent of geographic agglomeration in the study of digital divides (Pick and Sarkar, 2015) to formulate policy responses to achieve digital equity. Furthermore, a geographic approach allows the incorporation of policy impacts; for example, enrollments in the Affordable Connectivity Program (ACP) that was funded by the U.S. Congress during the COVID-19 pandemic to provide subsidized high-speed internet to households in need. In short, a geographic approach allows the proper contextualization of traditional sociodemographic, economic, and cultural influences on the technological, infrastructural, and policy dimensions of digital divides. Hence, a geographic approach is strongly recommended.

Next, considering the intersections between digital, social, cultural, economic, and demographic factors that interrelate, the following questions merit additional research and analysis to formulate meaningful policies to achieve digital equity and inclusion:

- What is the relationship between households without internet or personal vehicles? (Urban Observatory, 2024)
- What are the associations of English and Math proficiencies of K-12 students with high-speed broadband availability and coverage?
- What is the relationship between children who are in the care of their grandparents and children without broadband internet access? (Urban Observatory, 2024)
- What is the association between access to libraries (e.g., within a 10-minute walk) and the quality and extent of broadband coverage in urban, suburban, and rural areas?

These questions have implications for the built environment in both urban and rural parts of the nation. In many parts of the United States, the concept of livable, walkable, and equitable cities is gaining steam in the form of x-minute cities (where x = 15, 20, or 30 minutes). The idea is rooted in providing convenient, often walkable, access to places and services that citizens need on an everyday basis. In communities that continue to be afflicted by poor internet and broadband connectivity, citizens often rely on public libraries, community centers, and similar facilities to gain access to cyberspace. Hence, accessibility to such facilities, particularly for those who do not own personal cars or are unable to afford rideshares, has implications for the built environment.

The built environment of urban places has been analyzed extensively. It reveals inequities in the spatial distribution of income, wealth, education, and occupations within cities. These patterns can be theorized using models ranging from classic approaches such as concentric zones, sectors, and multiple nuclei to more recent models of structuralism and postmodernism (Greene and Pick, 2012). Added to this today are models of the spatial distribution of digital infrastructure. Smart cities is one recent approach in which contemporary technologies, including digital access and uses, are emphasized in the planning and management of city functions (Green, 2019). Smart cities are often characterized by digitally managed transport, buildings optimized through the internet of things, automation of city services, smart sensors, factories, and energy facilities, and technological city planning to achieve a more sustainable urban environment.

The narrowing of the digital divide in the United States can help to make smart cities realizable, in the sense that city inhabitants would share equitable access, usage, and skills to fully utilize the smart city systems being made available, as well as having the motivation to support smart city initiatives as citizens and voters. Although geographic differences in digital access and usage remain within and between cities, the narrowing of the digital divide over time among age groups, gender, and racial/ethnic categories (Pew Research Center, 2024) reveals the potential for the achievement of greater digital, as well as socioeconomic equity. According to recent Pew surveys, four out of every five U.S. homes have broadband connectivity in 2023 (Gelles-Watnick, 2024). This reveals the considerable capability of citizens to access (if not utilize) broadband and broadband-driven, smart-city infrastructure. This could include the buildout of affordable smart homes in which residents are able to interact with an array of household equipment using the internet and the reconceptualization of transportation infrastructure that is amenable to smart public transport systems and autonomous vehicles. Ultimately, the goal is the creation of a hybrid dual-space society in which individuals in their physical space can interact seamlessly with vital virtual spaces (work, school, and recreation), irrespective of their demographic and socioeconomic attributes and their place of residence or work. This, in turn, could improve equity of opportunity for many.

With these potential benefits in mind, trends in migration, technology infrastructure, and services call for more research exploration. One example is the shift in access to broadband across the urban-suburban-rural continuum as influenced by COVID-induced changes in broadband intensities of use (White and Glackin, 2023). As the built environment of most large cities has faced the challenges of overcrowding, traffic congestion, air pollution, large-scale manufacturing, and infectious diseases, there has been a push to migrate to the periphery. This tendency was exaggerated in many U.S. cities during the COVID-19 pandemic. Consequently, there was a significant movement of the population to the suburbs and the urban fringe (Ramani and Bloom, 2022). Some of this movement involved the relocation of highly broadband-skilled people to the urban periphery with "high-bandwidth internet and telecommuting [becoming] the catalyst for a return to lower urban densities and the suburban/village idyl" (White and Glackin, 2023). However, this movement challenges the built infrastructure of the urban fringe and rural small towns, with the necessity to provide new energy, telecommunications, and technical services. In terms of equity in the built environment, this scenario indicates both a geographic leveling of broadband infrastructure, skills, and use, yet also potentially an expansion of socioeconomic inequalities since migrants with high socioeconomic status might be disproportionally advantaged by new digital infrastructure and a lower cost of living (relative to larger, urban areas). It remains to be seen whether these trends will endure. Remote work is currently threatened by the return-to-office mandates as well as new investments to revitalize urban centers as hubs of commercial and social activities to restore their urban appeal (González-Leonardo, Rowe, and Fresolone-Caparrós, 2022).

Regarding trends in technical skills, as discussed earlier, broadband access does not necessarily equate to purposeful and meaningful use of the internet. Without the requisite skills, a digitally naïve individual will likely be unable to make purposeful use of the internet. Hence a two-pronged approach that blends the provision of access and skills is critical. Previously, in this chapter, the role of universities, military bases, and research institutions was highlighted in outlying rural counties that have high levels of digitalization but are in close proximity to other laggard counties with low to very low levels of digitalization. What roles can universities, research labs, and other similar institutions play in fostering access to the internet, enabling pathways for the development of digital knowledge and skills, and ensuring digital equity and inclusion? Additionally, how will the urban and rural built environments both facilitate and evolve in response to these changes? These, and other critical questions, must be pondered as we continue to strive toward a more mature, equitable, and inclusive digital society.

References

Access Broadband Dashboard. Available at https://www.census.gov/programs-surveys/community-resilience-estimates/partnerships/ntia/broadband-act.html 2023.

Adams, Paul C., and Barney Warf, eds. *Routledge Handbook of Media Geographies.* London, New York: Routledge, 2022.

Chen, Wenhong. "The implications of social capital for the digital divides in America". *The Information Society* 29, no. 1 (2013): 13–25.

FCC. *Inquiry Concerning Deployment of Advanced Telecommunications Capability to All Americans in a Reasonable and Timely Fashion*, 2024, Section 706 Report, Federal Communications Commission, Washington, D.C. Available at https://docs.fcc.gov/public/attachments/DOC-400675A1.pdf

Gallardo, Roberto, and Brian Whitacre. "An unexpected digital divide? A look at internet speeds and socioeconomic groups." *Telecommunications Policy* 48, no. 6 (2024): 102777.

Gelles-Watnick, Risa. "*American's Use of Mobile Technology and Home Broadband.*" Report, Pew Research Center, Jan. 31, 2024. Available at https://www.pewresearch.org/wp-content/uploads/sites/20/2024/01/PI_2024.01.31_Home-Broadband-Mobile-Use_FINAL.pdf

Goldberg, Rafi. "*New NTIA Data Show 13 Million More Internet Users in the U.S. in 2023 than 2021*". Blog, National Telecommunications and Information Administration, Washington, D.C. 2024. Available at https://www.ntia.gov/blog/2024/new-ntia-data-show-13-million-more-internet-users-us-2023-2021

González-Leonardo, Miguel, Francisco Rowe, and Alberto Fresolone-Caparrós. "Rural revival? The rise in internal migration to rural areas during the COVID-19 pandemic. Who moved and Where?" *Journal of Rural Studies* 96 (2022): 332–342.
Green, Ben. *The Smart Enough City: Putting Technology in Its Place to Reclaim Our Urban Future*. Cambridge, MA: MIT Press, 2019.
Greene, Richard P., and James B. Pick. *Exploring the Urban Community: A GIS Approach*. Upper Saddle River NJ: Prentice Hall, 2012.
Grishchenko, Natalia. "An inverted digital divide during Covid-19 pandemic? Evidence from a panel of EU countries." *Telematics and Informatics* 72 (2022): 101856.
Grubesic, Tony H., and Edward Helderop. "California's digital divide and the specter of data uncertainty for evaluating broadband coverage." *Telematics and Informatics* 71 (2022): 101837.
Hargittai, Eszter. 2002. "Second-level digital divide: Differences in people's online skills". *First Monday* 7 (4). https://doi.org/10.5210/fm.v7i4.942
Helsper, Ellen J., Alexander JAM van Deursen, and Rebecca Eynon. *Tangible outcomes of Internet use: From digital skills to tangible outcomes project report*. Oxford: Oxford Internet Institute. (2015). https://core.ac.uk/download/pdf/30671581.pdf
ITU. *ICT Statistics*. International Telecommunication Union. Available at https://www.itu.int/en/ITU-D/Statistics/Pages/stat/default.aspx 2024.
Khatiwada, Lila K., and Kenneth E. Pigg. "Internet service provision in the US counties: Is spatial pattern a function of demand?" *American Behavioral Scientist* 53, no. 9 (2010): 1326–1343.
Lythreatis, Sophie, Sanjay Kumar Singh, and Abdul-Nasser El-Kassar. "The digital divide: A review and future research agenda." *Technological Forecasting and Social Change* 175 (2022): 121359.
NTIA. *Exploring the Digital Nation: America's Emerging Online Experience*, 2013. National Telecommunications & Information Administration, Washington, D.C.
NTIA. *Exploring the Digital Nation: Embracing the Mobile Internet*, 2014. National Telecommunications & Information Administration, Washington, D.C.
NTIA. *NTIA Data Explorer*, 2024. National Telecommunications & Information Administration, https://www.ntia.gov/data/explorer#sel=internetUser&disp=map
OECD. "*Understanding the Digital Divide*", OECD Digital Economy Papers, No. 49, Organisation for Economic Cooperation and Development Publishing, Paris, 2001. https://doi.org/10.1787/236405667766
Pew Research Center. "*Internet, Broadband Fact Sheet*." Washington, D.C.: Pew Research Center. Jan. 31, 2024.
Pick, James B., and Avijit Sarkar. *The global digital divides: Explaining Change*. Berlin: Springer, 2015.
Pick, James, Fang Ren, and Avijit Sarkar. "Digital inequalities in China in 2020: Spatial and multivariate analysis." *Applied Sciences* 14, no. 13 (2024): 5385.
Pick, James, Avijit Sarkar, and Elizabeth Parrish. "The Latin American and Caribbean digital divide: A geospatial and multivariate analysis." *Information Technology for Development* 27, no. 2 (2021): 235–262.
Qureshi, Sajda. "As the global digital divide narrows, who is being left behind?." *Information Technology for Development* 18, no. 4 (2012): 277–280.
Rahman, M. M., T. Ara, and R. Chakma. (2022). Explaining geospatial variation in mobile phone ownership among rural women of Bangladesh: A multi-level and multidimensional approach. *Telecommunications Policy*, *46*(5), 102289.
Ramani, A., and N. Bloom. "*The Donut Effect of Covid-19 on Cities*." Working paper 28876, National Bureau of Economic Research, Cambridge, MA, December, 2022.
Sarkar, Avijit, and James B. Pick. "Pandemic Influences on Americans' Online Activities: A Spatial and Multivariate Analysis." In *Proceedings of the 51st Research Conference on Communication, Information and Internet Policy*. 2023, Washington, D.C. Available at SSRN: https://ssrn.com/abstract=4527462

Sarkar, Avijit, James B. Pick, and Tridev Raut. "Spatial Patterns of Purposeful Social Networking and Socioeconomic Influences: Towards a Mature, Inclusive Networked Society in the US." In *Proceedings of the 49th Research Conference on Communication, Information and Internet Policy*. 2021, Washington, D.C.

Sarkar, Avijit, James B. Pick, and Jessica Rosales. "Multivariate and geospatial analysis of technology utilization in US counties." *Telecommunications Policy* 47, no. 7 (2023): 102572.

Scheerder, Anique, Alexander Van Deursen, and Jan Van Dijk. "Determinants of internet skills, uses and outcomes. A systematic review of the second-and third-level digital divide." *Telematics and Informatics* 34, no. 8 (2017): 1607–1624.

Selwyn, Neil. "Reconsidering political and popular understandings of the digital divide." *New Media & Society* 6, no. 3 (2004): 341–362.

Setthasuravich, Prasongchai, and Hironori Kato. "Does the digital divide matter for short-term transportation policy outcomes? A spatial econometric analysis of Thailand." *Telematics and Informatics* 72 (2022): 101858.

Song, Zhouying, Chen Wang, and Luke Bergmann. "China's prefectural digital divide: Spatial analysis and multivariate determinants of ICT diffusion." *International Journal of Information Management* 52 (2020): 102072.

Urban Observatory. Atlas of internet connectivity. 2024. Available at https://urbanobservatory.maps.arcgis.com/apps/instant/portfolio/index.html?appid=45b65b0ef07c40efaf1e51ead94879e8

Van Deursen, Alexander JAM, and Jan AGM Van Dijk. *Digital Skills: Unlocking the Information Society*. New York NY: Springer, 2014.

Van Dijk, Jan. *The Digital Divide*. Medford MA: John Wiley & Sons, 2020.

Van Dijk, Jan, and Kenneth Hacker. "The digital divide as a complex and dynamic phenomenon." *The Information Society* 19, no. 4 (2003): 315–326.

Warf, Barney. *Global Geographies of the Internet*. New York, NY: Springer Science & Business Media, 2013.

Warf, Barney. *Geographies of the Internet*. Editor. New York, NY: Routledge, 2021.

White House. *Mapping the Digital Divide*. Council of Economic Advisors Issue Brief. 2015. Washington, D.C. Available at https://obamawhitehouse.archives.gov/sites/default/files/wh_digital_divide_issue_brief.pdf

White, Marcus, and Stephen Glackin. "Broadband-acre city: 'No traffic problem, no buffering'." *Architectural Design* 93, no. 1 (2023): 30–37.

15 Impact of the Urban Built Environment on Public Health during the COVID-19 Pandemic

Amit Patel

Introduction

The urban built environment that includes physical structures, open spaces, and infrastructure shapes our daily experiences and interactions in cities. Consequently, the urban built environment has a profound influence on our health and well-being in general, especially during public health emergencies such as the recent COVID-19 pandemic. For example, well-designed and well-maintained public spaces can promote physical activity, social connection, and community cohesion. Conversely, poorly designed or neglected spaces can lead to social isolation (Araya et al. 2006), decreased physical activity (Gordon-Larsen et al. 2006), and feelings of insecurity or fear (Hur and Nasar 2014).

During public health emergencies, poorly designed and neglected spaces exacerbate impacts, especially on marginalized populations. The COVID-19 pandemic has particularly highlighted existing inequalities in the urban built environment, especially through disparities in housing quality. People living in overcrowded neighborhoods or substandard housing are not only more likely to be exposed to the virus (von Seidlein et al. 2021) but also to experience severe illness if infected (Ahmad et al. 2020). Preventative measures such as frequent handwashing or social distancing are difficult to practice in places without access to clean water or overcrowded environments such as those found in slums of the Global South.

In this chapter, I discuss the ways in which inequality in the built environment has played a significant role in exacerbating COVID-19 inequality, particularly for marginalized communities in slums of the Global South. First, I provide a brief overview of the relationship between the urban built environment and public health outcomes. Next, I discuss the role of the urban built environment in the context of public health emergencies such as pandemics. I then discuss three of my collaborative studies on the COVID-19 pandemic and related public health emergencies with a specific focus on the urban built environment found in slums. Finally, I conclude with urban planning and policy recommendations to address these disparities in the urban built environment to not only make our cities more inclusive and livable for all but also healthier and more resilient in the face of future public health emergencies.

DOI: 10.4324/9781032631110-16

Urban Built Environment and Public Health Outcomes

Humanity has lived in communities for millennia and the built environment has been a defining feature of our settlements, from nomadic cultures to today's complex urban societies. The built environment is broadly defined as the human-made spaces in which people live, work, and recreate on a day-to-day basis (Roof and Oleru 2008). Since the Industrial Revolution, the built environment has become increasingly urban, providing shelter for large populations in a rapidly urbanizing world. To provide context, only a third of the world's population lived in cities in the 1950s; today, more than half of the global population lives in urban areas, and with increasing urbanization in the global South, it is projected to increase to around two-thirds by 2050 (United Nations, n.d.). To accommodate this urban population growth, cities have been expanding both in planned ways as well as informally (UN-Habitat 2003).

The urban built environment has recently been under scrutiny given its implications for public health outcomes, especially with many of the recent mental and physical health threats that urban populations have experienced. When studying public health outcomes, it is important to note that the urban built environment does not constitute only physical spaces defined by buildings, open spaces, and infrastructure but also include dynamics such as the movement of goods and people, intangible aspects of living in a city such as social networks and interactions between citizens, perceived safety, as well as characteristics such as types of land use and densities (Renalds, Smith, and Hale 2010; Leyden 2003). In that sense, the urban built environment is a broad concept that impacts nearly every aspect of the day-to-day lives of people living in cities.

There is much scientific evidence, especially in the field of urban health, documenting the health implications of urban built environments. One of the most widely cited relationships is between the physical urban built environment and the physical activity of urban residents (Abildso et al. 2007). Proximity to, and availability of, amenities such as shopping areas, parks, beaches, and trails leads to increased physical activity as numerous ecological psychology studies have found (Sallis, Bauman, and Pratt 1998). Corollary to this, perceived or actual lack of access to opportunities for physical activities leads to a more sedentary lifestyle (Spangler-Murphy et al. 2005). This health behavior (or lack thereof) in turn affects obesity rates and associated chronic diseases (Berke et al. 2007; Gordon-Larsen et al. 2006; Coutinho et al. 2023; Grafova 2008; Prados, Nicosia, and Datar 2023). Similarly, a poor urban built environment is associated with deleterious health outcomes such as increased depression (Galea et al. 2005; Mouratidis 2021) and unhealthy behaviors such as increased alcohol consumption (often related to the concentration of alcohol outlets in the neighborhood) (Bernstein et al. 2007; Trangenstein et al. 2020; Scott et al. 2020).

The Urban Built Environment and Epidemics

The urban built environment influences the prevention and containment of diseases, both chronic and infectious. For example, the built environment that promotes healthy behaviors such as physical activity and provides access to healthy foods can help prevent chronic diseases such as obesity and associated diseases such as diabetes and heart disease (Pinter-Wollman, Jelić, and Wells 2018). However, spatial segregation in cities by race, class, and ethnicity creates an uneven landscape of health (often through disinvestment in poorer areas), leading to unequal health outcomes between various socioeconomic groups. Similarly, infectious diseases such as influenza and tuberculosis spread much faster in urban environments because the pathogens rely on contact and proximity that the highly dense areas of cities readily provide (Alirol et al. 2011). In addition to high density, cities are also hubs for rural–urban and urban–urban migration that provide favorable grounds for germs to spread as they travel with migrants (Bella et al. 1980). While cities provide economic opportunities and reduce poverty for rural migrants, they are also characterized by inequalities in economic, social, and living conditions (Brockerhoff and Brennan 1998; Nijman and Wei 2020). For instance, in the U.S., undocumented immigrants face housing affordability problems more often than legal immigrants or the native population. Similarly, in Chinese cities, registered residents (*hukou* permit holders) receive benefits from the government's affordable housing policies while migrants without registration are usually excluded (Huang 2012). While migration itself could serve as a livelihood strategy to escape rural poverty, the health dividend of migration to cities is therefore often offset by a health burden among migrant groups through such exclusionary policies and practices (Bhan and Patel 2023).

The urban built environment plays a significant role in how infectious diseases spread as their diffusion is directly tied to urban mobility as well as human–human and human–environment interactions. The attributes of the urban built environment including physical as well as social configurations play a major role in the transmission of disease. For instance, Aguilar et al. (2022) showed that in U.S. cities where the movement of residents was concentrated between mobility hotspots, there was a relatively rapid spread of COVID-19, while more sprawling cities with multiple nodes of activity saw a slower spread of the disease. Non-pharmaceutical interventions such as restrictions on mobility, however, were more effective at curbing the disease in cities with concentrated mobility hotspots. Similarly, a metanalysis of studies on African cities reported that high population density, poor housing, and poor sanitation infrastructure are directly related to the spread of infectious diseases and, because these risk factors are uneven within a city, the disease spread is also uneven (Boyce, Katz, and Standley 2019). Importantly, these spatial and aspatial configurations are diverse for individuals and hence transmission is not equal across various socioeconomic groups (Salathé and Jones 2010).

The non-pharmacological measures to prevent the spread of diseases are also a function of the urban built environment (Barro 2022), especially in slowing the spread of disease as observed during the Great Influenza Pandemic of 1918. However, non-pharmacological measures such as quarantine for infected individuals within a building or neighborhood largely depend on the configuration of the built environment, especially density and the self-sufficiency of neighborhoods for essential supplies. For instance, in the self-contained but very dense slum of Dharavi in Mumbai, India, several shops and residential buildings were sealed, and many residents were quarantined when the first case of COVID-19 was detected (Biswas 2020). However, such geographic quarantines did not fully contain the spread of disease in Mumbai, which is characterized by particularly high population density both within and outside of slums. Similarly, providing safe transportation options not only require alternatives to crowded transport systems but also frequent cleaning of vehicles and surfaces, temperature checks for staff and even passengers, and increased air ventilation that many cities and governments introduced during the COVID-19 pandemic (Tirachini and Cats 2020; Gkiotsalitis and Cats 2021). Many cities reduced the capacity of public trains and buses to lessen crowding and help curb the spread of disease. The city of Denver, Colorado, for example, instituted a capacity limit of 15 passengers per city bus and 30 per rail car to enable social distancing. Surgical masks were also offered to riders in Detroit to reduce exposure to bus operators (Hughes 2020).

COVID-19 and the Urban Built Environment in Slums

While there is no dearth of literature on the urban built environment and its relationship with chronic and infectious disease epidemics and pandemics in the developed world, evidence of inequalities in the urban built environment in the developing world is more limited. This gap in knowledge became evident during the recent COVID-19 pandemic, and especially in the context of informal settlements or slums (Corburn et al. 2020), which are home to over a billion people globally or one out of every three urban residents (UN, n.d.).

One of the most prominent manifestations of intra-city inequality is in large cities of the Global South where slums coexist alongside formal developments. There is often a stark contrast between these settlements that is evident in housing quality, access to basic infrastructure such as water and sanitation, paved streets, lighting, and other neighborhood amenities (UN-Habitat 2003). The COVID-19 pandemic posed a challenge to slum populations especially because the urban built environment of slums made them particularly vulnerable to the rapid spread of disease and standard public health measures and non-pharmacological interventions were often counterproductive and burdensome (Patel 2020; Iwuoha and Aniche 2020).

Within the field of urban health, slum health has emerged as an important research area (Corburn and Riley 2016) with a special emphasis on slum health equity (Corburn 2016). The recent COVID-19 pandemic brought specific vulnerabilities and risks for slum dwellers around the globe to the fore that sparked a strong call for practice and policy that is cognizant of the economic and social realities of slums (Corburn et al. 2020). Scholars and practitioners emphasized that challenges associated with containment, mitigation, and control in low-resource settings of informal settlements demanded special attention. Most basic needs such as water, sanitation, waste management, and adequate housing are in short supply or nonexistent, which makes slum populations particularly susceptible to infectious diseases (Lilford et al. 2017). Consequently, top-down approaches that ignore these realities make impractical recommendations such as social distancing that are highly unlikely to be effective in curbing the spread of disease (Corburn et al. 2020).

Slums provide informal housing but are also places for substantial economic activities and thus any response to COVID-19 should also consider balancing economic well-being with curbing the spread of disease. This was particularly challenging because many national governments took drastic measures such as lockdowns that ignored the economic reality of cities in the Global South. In the remainder of this section, I will discuss three of my studies on COVID-19 with a specific focus on the built environment of slums. This work suggests that inequalities in cities of the Global South require specific attention to manage public health emergencies.

The COVID-19 pandemic brought renewed attention to the lack of urban planning and the challenges slums face that have historically been ignored. Patel (2020) showed that public health measures in the face of the pandemic often ignored the realities of the physical built environment in slums and thus could be ineffective or counterproductive. Using data from Indian cities, the study highlighted that one of the most prescribed preventive measures, social distancing, posed a significant challenge in practice. For instance, Dharavi, a large slum in central Mumbai, is 12 times denser than Mumbai on average, and roughly 30 times denser than New York City. Furthermore, the low-rise high-density built environment contains very small dwelling units that are tightly spaced with narrow lanes, making it extremely difficult to maintain social distancing. In addition, most slum-dwellers rely on shared water and sanitation facilities, bringing them in contact with each other frequently throughout the day.

Similarly, family members of confirmed COVID-19 patients were expected to home quarantine during the pandemic with an expectation that the patient would stay in a separate room for 14 days without coming into contact with family members. Strikingly, 45% of dwelling units in slums in India had only a single room, making it impossible to home quarantine. Highly dense environments with unique living conditions and socioeconomic characteristics make one-size-fits-all measures nearly ineffective (Patel 2020).

Frequent handwashing was yet another public health measure instituted to help curb the pandemic. However, lack of access to clean water and poor sanitation in slums made it extremely challenging to follow this guideline. Frequent handwashing at a community facility could in fact increase the risk of disease transmission (Caruso and Freeman 2020; Hasan et al. 2021; Kotwal et al. 2020).

As discussed earlier, the urban built environment also includes dynamics such as the movement of people, their interactions, and social networks. Complete lockdowns on short notice during the pandemic left many migrant workers in Indian cities out of work and stranded in their destination cities.

Many of them considered walking hundreds of miles in desperation, and some likely carried the disease back to their rural origins. Patel and Shah (2020) discussed the characteristics of the urban built environment and highlighted that cities are highly porous with migrants originating from a vast urban–rural network. Any attempt to lockdown cities with an assumption that they are constrained within their physical boundaries will be ignorant of the porous nature of urban built environments. Slums in that sense are not limited to the small physical area that they occupy, but rather they are nodes that connect to numerous villages around the country from where many of the slum residents originate. Such a network view of the urban built environment largely negates public health emergency measures that rely on geographic quarantines. The authors suggest, however, that networks within slums and between slums and non-slums are important characteristics of the urban built environment that could be leveraged to address public health emergencies as Corburn et al. (2020) have also advocated. For instance, a partnership between the Ahmedabad (India) Municipal Corporation (AMC) and Self-Employed Women's Organization (SEWA) created an innovative program called "Vegetables on Wheels" that ensured essential supplies to everyone in the city of Ahmedabad, India during the pandemic while also providing a livelihood for slum dwellers (Chen 2020). In Bengaluru, Swiggy, a food delivery service similar to Uber Eats, utilized its solid understanding of food supply networks to cater half a million meals a day to vulnerable populations during lockdown in India (PTI 2020).

This view of the urban built environment as a network has its origins in complexity science, which has informed a new science of cities (Batty 2013) that provides a radically different view of urban places. In this view, cities are not simply places in space, but rather systems of networks and flows. Batty (2013) suggests that if we want to understand cities and how they function, we must first understand them as flows within networks of goods, people, and information. These networks are further composed of nodes such as markets and linkages such as public transit systems. In their think piece, Patel and Shah (2020) propose that the new science of cities (Batty 2013), combined with the subfield of slum health (Corburn and Riley 2016; Riley et al. 2007) within the broader field of urban health, could reveal more effective ways of dealing with public health emergencies such as the COVID-19 pandemic. This is particularly relevant for cities in the Global South which are often characterized by inequalities manifested in the built environment of slums. For example, a network view of slums would have explicitly recognized that migrants in slums are not only residents of a specific slum but are also part of a network of destinations and origins connected by regional transport systems. Such a porous view of the city suggests the need to arrange safe transportation between migrants' place of origin and their destination cities prior to implementing a city-level lockdown, as well as preparing self-quarantine facilities for their return. This network view of slums and cities breaks from the traditional view of slums as simply a place within a city. The traditional view has generally resulted in failed attempts at geographic quarantines such as in Dharavi in Mumbai as discussed earlier.

The new science of cities also explicitly recognizes agency and human behavior in response to policies. As was seen in many Indian cities, including New Delhi, migrants walked hundreds of kilometers to care for their dependents back home as well as to escape the precarity of their destination city when they were deprived of livelihood and sustenance (Abi-Habib and Yasir 2020). Subsequently, the government of India introduced a special train service to take migrant workers back to their origins (i.e., *Shramik* Trains), but it was too late for many (Patel and Shah 2020). A network view that envisages destination cities and origin villages as nodes connected by migration flows as the linkages could easily anticipate safe transportation requirements for migrants as part of a comprehensive public health emergency response.

This new way of looking at slums and cities could also inform how we model disease transmission and assess the effectiveness of our proposed public health measures. For instance, French et al. (2024) modeled a slum in Ahmedabad, India to show how varying densities within a slum influence the effectiveness of various public health measures such as mandatory masks and geographic quarantine. The authors showed that partial lockdowns without face mask mandates cause a significant increase in disease spread in high-density areas within the slum. Such explicit recognition of human–human and human–environment interactions that are characterized by diversity and heterogeneity of people as well as spaces within a slum gives us a better understanding of reality and prevents us from the false confidence that may come with centrally planned urban health measures.

Opportunities for Urban Planning and Policy

Urban planning has historically been a direct response to unhygienic living conditions and health concerns (Sloane 2006). Modern architecture and planning practices have improved living conditions significantly over the last century. However, modern cities are now recognized as complex systems characterized by nonlinear, dynamic, self-organized growth that often defies urban planning. Recognizing cities as complex systems is the first step toward better urban health outcomes (Gatzweiler et al. 2017). In that sense, informal settlements are a bottom-up response to planning and policy failures of centrally orchestrated utopian visions that are unlikely to materialize. Complex systems, however, are extremely resilient and carry significant strengths that could be leveraged to address public health challenges including pandemics and other emergencies. For example, in the favelas of Sao Paulo, a slum community hired their own doctor to fight the pandemic in the absence of any centralized urban health response (Andrade 2020).

In this final section, I build on the notion of urban planning as a preventive medicine that Corburn (2015) has proposed to improve urban health outcomes in the face of extreme urban inequalities. The COVID-19 pandemic has been studied in detail from its onset and remains a subject of scientific investigation. I posit that we now have access to ample evidence documenting the successes and failures of urban planning and policy measures in the context of COVID-19 from multiple disciplines. Cities could plan their built environment in such a way that makes them safer, more resilient, and more equitable during pandemics, especially for marginalized populations living in slums. Toward this goal, cities in Global South could: (1) consider slums and non-slums as part of one integral physical environment that is planned to ensure inclusion; (2) develop equitable zoning policies to increase serviced land supply and facilitate creation of affordable housing stock to reduce overcrowding; (3) work with existing slum communities to improve the built environment by integrating them into the city's water, sanitation, solid waste management, transportation, healthcare, and education infrastructure; (4) recognize and address the unmet needs of the most vulnerable groups within slums by investing in amenities such as playgrounds for children, parks for senior adults, and lodging and dining facilities for single migrants, and (5) upgrade the connectivity between slums and non-slums to improve the flow of people and essential supplies during public health emergencies.

The new science of cities could also be helpful in developing urban planning and policy measures that recognize urban areas as complex systems and slums as complex systems embedded within. In addition, cities themselves are part of large regional and global complex systems. Such interconnectedness with a variety of networks of goods, people, and information, inevitably affects the transmission of diseases, which is nonlinear and requires preventive measures that are adaptive and dynamic. Such measures could include, for instance, an integrated system of urban governance that can rapidly detect infected individuals, take timely lockdown and social distancing actions (Duggal et al. 2022), and provide economic and social support to vulnerable populations as, for example, seen in Vietnam's Ho Chi Minh city (Thoi 2020). Cities and nations could also develop emergency plans that are informed by the lessons learned from the current pandemic as Singapore, Taiwan, and South Korea have done. These countries could better respond to COVID-19 because they had developed such plans based on their experience with the previous SARS and H1N1 pandemics (Sharifi and Khavarian-Garmsir 2020). Immediate activation of an existing Municipal Emergency Plan that is often developed for natural disasters is another adaptive and dynamic response as seen in Setubal, Portugal (Santos et al. 2020).

Urban planning and policies can urgently focus on highly dense urban environments such as slums and implement measures that could reduce the negative effects of such densities and lack of basic services. For example, slum upgrading programs can improve the pedestrian and vehicular flow in a slum by paving the streets and removing the bottlenecks in narrow streets for better emergency response. Planners and policymakers can also emphasize to the community the often-significant implications the urban built environment has for health emergencies and work with them to prioritize the provision of water, sanitation, and hygiene (WASH) facilities. There are many examples of local officials successfully engaging with communities to carry out participatory improvements such as the Slum Networking Project undertaken by the city of Ahmedabad to improve living conditions through physical upgrading of the built environment (e.g., paved roads, street lighting, toilets, water supply) (Das and Takahashi 2009).

As Patel and Shah (2020) have argued, we need to integrate principles of slum health with the new science of cities to inform both short-term pandemic response and enable long-term planning to make our cities healthier, resilient, and more equitable for all. Towards this goal, we need to recognize that the community is an integral part of urban planning (Weimann et al. 2020; Corburn et al. 2019), and the emergence of informal settlements is an indication of formal planning failure that needs to be addressed with residents, not for them. In that sense, any top-down planning in the long term or public health measures during epidemic outbreaks in the short term are less likely to be effective in achieving equitable outcomes. Similarly, intracity and intercity networks must be recognized in urban planning processes as opposed to treating cities as closed physical systems. The urban built environment is larger than the sum of its parts, comprised of a variety of overlapping and interacting networks that include both slums and non-slums. These networks must be recognized and leveraged in planning and public health measures as demonstrated by the innovative approaches taken by the city of Ahmedabad and Bengaluru in India or Sao Paulo's favela during the COVID-19 pandemic discussed earlier. One possibility, as proposed by Corburn (2020), is to create emergency planning committees at the slum/neighborhood level that are networked across cities and the country and could respond to public health emergencies faster and more effectively. Corburn (2020) also suggests that plans for mobility and healthcare include the identification of evacuation routes within slums that avoid narrow lanes and provide access to ambulances and other emergency vehicles that connect to healthcare facilities outside slums.

This shift in paradigm toward an interconnected and complex urban built environment will inevitably require us to pursue scholarship with a different worldview and draw practice and policy implications that are both transformative and inclusionary. Such a conceptualization makes it clear that a local case of a new disease is not necessarily an isolated episode but has the capacity to spread locally, regionally, and globally in little time. In the same vein, such a system is also capable of coordinating local efforts to help curb global pandemics. One way to control the pandemic is to develop a decentralized information system that cuts across boundaries and collects spatially explicit data at a variety of geographic scales to inform the spread of diseases within and between neighborhoods and cities. The pandemic and the mobility restrictions through lockdowns and geographic quarantines also suggest that cities in the future will need to develop more agile and responsive transportation systems that can adapt quickly and take action to meet the needs of residents as well as limit the spread of disease.

Modern urban planning was driven in large part by public health concerns, particularly the threat of infectious diseases. In the mid-19th century, John Snow famously mapped cases of Cholera in London to identify the source as a single water pump, an exercise that has made mapping an essential part of urban planning since (Ruths 2009). It is time that new epidemiological and computational modeling techniques be widely implemented in the urban planning process so that we may design cities that are more resilient and less susceptible to the rapid spread of diseases. Additionally, slums and cities in the Global South will not be prepared to respond to public health emergencies such as COVID-19 unless we accept the reality of their urban built environments and address their vulnerabilities. A well-prepared city of the future will: (1) look both within and beyond its boundaries with a network perspective; (2) build on lessons learned from cities around the world during this most recent pandemic, as well as infectious disease epidemics of the past; and (3) address inequalities and slum health concerns explicitly to make our cities more resilient, healthier, and more livable for all.

References

Abi-Habib, Maria, and Sameer Yasir. 2020. "India's Coronavirus Lockdown Leaves Vast Numbers Stranded and Hungry." *The New York Times*, March 29, 2020, sec. World. https://www.nytimes.com/2020/03/29/world/asia/coronavirus-india-migrants.html

Abildso, Christiaan G., Sam Zizzi, Laurie C. Abildso, Jenessa C. Steele, and Paul M. Gordon. 2007. "Built Environment and Psychosocial Factors Associated with Trail Proximity and Use." *American Journal of Health Behavior* 31 (4): 374–83. https://doi.org/10.5993/AJHB.31.4.4

Aguilar, Javier, Aleix Bassolas, Gourab Ghoshal, Surendra Hazarie, Alec Kirkley, Mattia Mazzoli, Sandro Meloni, et al. 2022. "Impact of Urban Structure on Infectious Disease Spreading." *Scientific Reports* 12 (1): 3816. https://doi.org/10.1038/s41598-022-06720-8

Ahmad, Khansa, Sebhat Erqou, Nishant Shah, Umair Nazir, Alan R. Morrison, Gaurav Choudhary, and Wu Wen-Chih. 2020. "Association of Poor Housing Conditions with COVID-19 Incidence and Mortality across US Counties." *PloS one* 15 (11): e0241327. https://doi.org/10.1371/journal.pone.0241327

Alirol, Emilie, Laurent Getaz, Beat Stoll, François Chappuis, and Louis Loutan. 2011. "Urbanisation and Infectious Diseases in a Globalised World." *The Lancet Infectious Diseases* 11 (2): 131–41. https://doi.org/10.1016/S1473-3099(10)70223-1

Andrade, Rodrigo de Oliveira. 2020. "The Brazilian Slums Hiring Their Own Doctors to Fight Covid-19." *BMJ* 369 (April). https://doi.org/10.1136/bmj.m1597

Araya, Ricardo, Frank Dunstan, Rebecca Playle, Hollie Thomas, Stephen Palmer, and Glyn Lewis. 2006. "Perceptions of Social Capital and the Built Environment and Mental Health." *Social Science & Medicine* 62 (12): 3072–83. https://doi.org/10.1016/j.socscimed.2005.11.037

Barro, Robert J. 2022. "Non-Pharmaceutical Interventions and Mortality in U.S. Cities During the Great Influenza Pandemic, 1918–1919." *Research in Economics* 76 (2): 93–106. https://doi.org/10.1016/j.rie.2022.06.001

Batty, Michael. 2013. *The New Science of Cities*. MIT Press.

Bella, Hassan, T. F. C. de Marshall, A. H. S. Omer, and J. P. Vaughan. 1980. "Migrant Workers and Schistosomiasis in the Gezira, Sudan." *Transactions of the Royal Society of Tropical Medicine and Hygiene* 74 (1): 36–39. https://doi.org/10.1016/0035-9203(80)90008-5

Berke, Ethan M., Thomas D. Koepsell, Anne Vernez Moudon, Richard E. Hoskins, and Eric B. Larson. 2007. "Association of the Built Environment with Physical Activity and Obesity in Older Persons." *American Journal of Public Health* 97 (3): 486–92. https://doi.org/10.2105/AJPH.2006.085837

Bernstein, Kyle T., Sandro Galea, Jennifer Ahern, Melissa Tracy, and David Vlahov. 2007. "The Built Environment and Alcohol Consumption in Urban Neighborhoods." *Drug and Alcohol Dependence* 91 (2): 244–52. https://doi.org/10.1016/j.drugalcdep.2007.06.006

Bhan, Tanushree, and Amit Patel. 2023. "Migration and Health: Exploring the Effect of Destinations on the Health Outcomes among Rural-Urban Migrants in India." *Social Science & Medicine* 331 (August):116079. https://doi.org/10.1016/j.socscimed.2023.116079

Biswas, Soutik. 2020. "Inside the Race to Halt the Virus in a Vast Indian Slum." *BBC News*, April 6, 2020, sec. India. https://www.bbc.com/news/world-asia-india-52159986

Boyce, Matthew R., Rebecca Katz, and Claire J. Standley. 2019. "Risk Factors for Infectious Diseases in Urban Environments of Sub-Saharan Africa: A Systematic Review and Critical Appraisal of Evidence." *Tropical Medicine and Infectious Disease* 4 (4): 123. https://doi.org/10.3390/tropicalmed4040123

Brockerhoff, Martin, and Ellen Brennan. 1998. "The Poverty of Cities in Developing Regions." *Population and Development Review* 24 (1): 75–114. https://doi.org/10.2307/2808123

Caruso, Bethany A, and Matthew C Freeman. 2020. "Shared Sanitation and the Spread of COVID-19: Risks and next Steps." *The Lancet: Planetary Health* 4 (5): e173. https://doi.org/10.1016/S2542-5196(20)30086-3

Chen, Marty Alter. 2020. "Vegetables on Wheels in Ahmedabad, India: SEWA Partners with Municipality to Ensure Food Access during Lockdown | WIEGO." May 5, 2020. https://www.wiego.org/blog/vegetables-wheels-ahmedabad-india-sewa-partners-municipality-ensure-food-access-during

Corburn, Jason. 2015. "City Planning as Preventive Medicine." *Preventive Medicine* 77 (August):48–51. https://doi.org/10.1016/j.ypmed.2015.04.022

Corburn, Jason. 2016. "Frameworks for Urban Slum Health Equity." In *Slum Health: From the Cell to the Street*, edited by Jason Corburn and Lee Riley, 51–69. Univ of California Press.

Corburn, Jason, and Lee Riley. 2016. *Slum Health: From the Cell to the Street*. Univ of California Press.

Corburn, Jason, Marisa Ruiz Asari, Jorge Pérez Jamarillo, and Aníbal Gaviria. 2019. "The Transformation of Medellin into a 'City for Life:' Insights for Healthy Cities." *Cities & Health* 4 (1): 13–24. https://doi.org/10.1080/23748834.2019.1592735

Corburn, Jason, David Vlahov, Blessing Mberu, Lee Riley, Waleska Teixeira Caiaffa, Sabina Faiz Rashid, Albert Ko, et al. 2020. "Slum Health: Arresting COVID-19 and Improving Well-Being in Urban Informal Settlements." *Journal of Urban Health*. https://doi.org/10.1007/s11524-020-00438-6

Coutinho, Sílvia R., Oddbjørn Klomsten Andersen, Nanna Lien, and Mekdes K. Gebremariam. 2023. "Neighborhood Deprivation, Built Environment, and Overweight in Adolescents in the City of Oslo." *BMC Public Health* 23 (1): 812. https://doi.org/10.1186/s12889-023-15261-2

Das, Ashok K., and Lois M. Takahashi. 2009. "Evolving Institutional Arrangements, Scaling Up, and Sustainability: Emerging Issues in Participatory Slum Upgrading in Ahmedabad, India." *Journal of Planning Education and Research* 29 (2): 213–32. https://doi.org/10.1177/0739456X09348613

Duggal, Chetna, Sukanya Ray, Rajani Konantambigi, and Arushi Kothari. 2022. "The Nowhere People: Lived Experiences of Migrant Workers during Covid-19 in India." *Current Psychology (New Brunswick, N.j.)* 41 (11): 8214–23. https://doi.org/10.1007/s12144-021-02220-6

French, Molly, Amit Patel, Abid Qureshi, Deepak Saxena, and Raja Sengupta. 2024. "Agent-Based Modeling of COVID-19 Transmission: A Case Study of Housing Densities in Sankalitnagar, Ahmedabad." *ISPRS International Journal of Geo-Information* 13 (6): 208. https://doi.org/10.3390/ijgi13060208

Galea, Sandro, Jennifer Ahern, Sasha Rudenstine, Zachary Wallace, and David Vlahov. 2005. "Urban Built Environment and Depression: A Multilevel Analysis." *Journal of Epidemiology & Community Health* 59 (10): 822–27. https://doi.org/10.1136/jech.2005.033084

Gatzweiler, Franz W., Stefan Reis, Yi Zhang, and Saroj Jayasinghe. 2017. "Lessons from Complexity Science for Urban Health and Well-Being." *Cities & Health* 1 (2): 210–23. https://doi.org/10.1080/23748834.2018.1448551

Gkiotsalitis, Konstantinos, and Oded Cats. 2021. "Public Transport Planning Adaption under the COVID-19 Pandemic Crisis: Literature Review of Research Needs and Directions." *Transport Reviews* 41 (3): 374–92. https://doi.org/10.1080/01441647.2020.1857886

Gordon-Larsen, Penny, Melissa C. Nelson, Phil Page, and Barry M. Popkin. 2006. "Inequality in the Built Environment Underlies Key Health Disparities in Physical Activity and Obesity." *Pediatrics* 117 (2): 417–24. https://doi.org/10.1542/peds.2005-0058

Grafova, Irina B. 2008. "Overweight Children: Assessing the Contribution of the Built Environment." *Preventive Medicine* 47 (3): 304–8. https://doi.org/10.1016/j.ypmed.2008.04.012

Hasan, Shaikh Mehdi, Susmita Das, Syed Manzoor Ahmed Hanifi, Sohana Shafique, Sabrina Rasheed, and Daniel D. Reidpath. 2021. "A Place-Based Analysis of COVID-19 Risk Factors in Bangladesh Urban Slums: A Secondary Analysis of World Bank Microdata." *BMC Public Health* 21 (1): 502. https://doi.org/10.1186/s12889-021-10230-z

Huang, Youqin. 2012. "Low-Income Housing in Chinese Cities: Policies and Practices." *The China Quarterly* 212 (December):941–64. https://doi.org/10.1017/S0305741012001270

Hughes, Trevor. 2020. "Poor, Essential and on the Bus: Coronavirus Is Putting Public Transportation Riders at Risk." *USA TODAY*, April 14, 2020. https://www.usatoday.com/story/news/nation/2020/04/14/public-transportation-users-risk-coronavirus-spreads-across-us/2979779001/

Hur, Misun, and Jack L. Nasar. 2014. "Physical Upkeep, Perceived Upkeep, Fear of Crime and Neighborhood Satisfaction." *Journal of Environmental Psychology* 38 (June):186–94. https://doi.org/10.1016/j.jenvp.2014.02.001

Iwuoha, Victor Chidubem, and Ernest Toochi Aniche. 2020. "Covid-19 Lockdown and Physical Distancing Policies Are Elitist: Towards an Indigenous (Afro-Centred) Approach to Containing the Pandemic in Sub-Urban Slums in Nigeria." *Local Environment* 25 (8): 631–40. https://doi.org/10.1080/13549839.2020.1801618

Kotwal, Nandita, Rebecca J Gilsdorf, Nishtha Mehta, and Martin Gambrill. 2020. "To Share or Not to Share Sanitation Facilities: A Coronavirus Conundrum." World Bank Blogs. July 30, 2020. https://blogs.worldbank.org/en/water/share-or-not-share-sanitation-facilities-coronavirus-conundrum

Leyden, Kevin M. 2003. "Social Capital and the Built Environment: The Importance of Walkable Neighborhoods." *American Journal of Public Health* 93 (9): 1546–51. https://doi.org/10.2105/AJPH.93.9.1546

Lilford, Richard J., Oyinlola Oyebode, David Satterthwaite, G. J. Melendez-Torres, Yen-Fu Chen, Blessing Mberu, Samuel I. Watson, et al. 2017. "Improving the Health and Welfare of People Who Live in Slums." *The Lancet* 389 (10068): 559–70. https://doi.org/10.1016/S0140-6736(16)31848-7

Mouratidis, Kostas. 2021. "Urban Planning and Quality of Life: A Review of Pathways Linking the Built Environment to Subjective Well-Being." *Cities* 115 (August):103229. https://doi.org/10.1016/j.cities.2021.103229

Nijman, Jan, and Yehua Dennis Wei. 2020. "Urban Inequalities in the 21st Century Economy." *Applied Geography* 117 (April):102188. https://doi.org/10.1016/j.apgeog.2020.102188

Patel, Amit. 2020. "Preventing COVID-19 Amid Public Health and Urban Planning Failures in Slums of Indian Cities." *World Medical & Health Policy* 12 (3): 266–73. https://doi.org/10.1002/wmh3.351

Patel, Amit, and Phoram Shah. 2020. "Rethinking Slums, Cities, and Urban Planning: Lessons from the COVID-19 Pandemic." *Cities & Health* 5 (sup1): S145–S147. https://doi.org/10.1080/23748834.2020.1790252

Pinter-Wollman, Noa, Andrea Jelić, and Nancy M. Wells. 2018. "The Impact of the Built Environment on Health Behaviours and Disease Transmission in Social Systems." *Philosophical Transactions of the Royal Society B: Biological Sciences* 373 (1753): 20170245. https://doi.org/10.1098/rstb.2017.0245

Prados, María J., Nancy Nicosia, and Ashlesha Datar. 2023. "Impact of Built, Social, and Economic Environments on Adolescent Obesity and Related Health Behaviors." *Obesity* 31 (4): 1085–94. https://doi.org/10.1002/oby.23682

PTI. 2020. "Swiggy Plans to Serve 5 Lakh Meals Daily to the Needy amid Coronavirus Lockdown | Deccan Herald." April 2, 2020. https://www.deccanherald.com/business/business-news/swiggy-plans-to-serve-5-lakh-meals-daily-to-the-needy-amid-coronavirus-lockdown-820599.html

Renalds, Arlene, Tracey H. Smith, and Patty J. Hale. 2010. "A Systematic Review of Built Environment and Health." *Family & Community Health* 33 (1): 68. https://doi.org/10.1097/FCH.0b013e3181c4e2e5

Riley, Lee W., Albert I. Ko, Alon Unger, and Mitermayer G. Reis. 2007. "Slum Health: Diseases of Neglected Populations." *BMC International Health and Human Rights* 7 (March):2. https://doi.org/10.1186/1472-698X-7-2

Roof, Karen, and Ngozi Oleru. 2008. "Public Health: Seattle and King County's Push for the Built Environment." *Journal of Environmental Health* 71 (1): 24–27.

Ruths, Mitali Banerjee. 2009. "The Lesson of John Snow and the Broad Street Pump." *AMA Journal of Ethics* 11 (6): 470–72. https://doi.org/10.1001/virtualmentor.2009.11.6.mhst1-0906

Salathé, Marcel, and James H. Jones. 2010. "Dynamics and Control of Diseases in Networks with Community Structure." *PLOS Computational Biology* 6 (4): e1000736. https://doi.org/10.1371/journal.pcbi.1000736

Sallis, James F, Adrian Bauman, and Michael Pratt. 1998. "Environmental and Policy Interventions to Promote Physical Activity a." *American Journal of Preventive Medicine* 15 (4): 379–97. https://doi.org/10.1016/S0749-3797(98)00076-2

Santos, Angela, Nuno Sousa, Horst Kremers, and José Luís Bucho. 2020. "Building Resilient Urban Communities: The Case Study of Setubal Municipality, Portugal." *Geosciences* 10 (6): 243. https://doi.org/10.3390/geosciences10060243

Scott, Jennifer, Denise Danos, Robert Collins, Neal Simonsen, Claudia Leonardi, Richard Scribner, and Denise Herd. 2020. "Structural Racism in the Built Environment: Segregation and the Overconcentration of Alcohol Outlets." *Health & Place* 64 (July):102385. https://doi.org/10.1016/j.healthplace.2020.102385

von Seidlein, Lorenz, Graham Alabaster, Jacqueline Deen, and Jakob Knudsen. 2021. "Crowding Has Consequences: Prevention and Management of COVID-19 in Informal Urban Settlements." *Building and Environment* 188 (January):107472. https://doi.org/10.1016/j.buildenv.2020.107472

Sharifi, Ayyoob, and Amir Reza Khavarian-Garmsir. 2020. "The COVID-19 Pandemic: Impacts on Cities and Major Lessons for Urban Planning, Design, and Management." *Science of The Total Environment* 749 (December):142391. https://doi.org/10.1016/j.scitotenv.2020.142391

Sloane, David Charles. 2006. "Longer View: From Congestion to Sprawl: Planning and Health in Historical Context." *Journal of the American Planning Association* 72 (1): 10–18. https://doi.org/10.1080/01944360608976720

Spangler-Murphy, Emily, Debra A. Krummel, Nicole Morrison, and Paul M. Gordon. 2005. "Environmental Perceptions Related to Physical Activity in High- and Low-Risk Counties." *Health Promotion Practice* 6 (1): 57–63. https://doi.org/10.1177/1524839903260157

Thoi, Pham Thanh. 2020. "Ho Chi Minh City- the Front Line against COVID-19 in Vietnam." *City & Society* 32 (2). https://doi.org/10.1111/ciso.12284

Tirachini, Alejandro, and Oded Cats. 2020. "COVID-19 and Public Transportation: Current Assessment, Prospects, and Research Needs." *Journal of Public Transportation* 22 (1). https://doi.org/10.5038/2375-0901.22.1.1

Trangenstein, Pamela J., Claire Gray, Matthew E. Rossheim, Richard Sadler, and David H. Jernigan. 2020. "Alcohol Outlet Clusters and Population Disparities." *Journal of Urban Health* 97 (1): 123–36. https://doi.org/10.1007/s11524-019-00372-2

UN. n.d. "SDG Indicators." Accessed June 26, 2024. https://unstats.un.org/sdgs/report/2019/goal-11/

UN-Habitat. 2003. "*The Challenge of Slums: Global Report on Human Settlements 2003*." Earthscan.

United Nations. n.d. "Urbanization | Population Division." United Nations | Population Division. Accessed June 18, 2024. https://www.un.org/development/desa/pd/content/urbanization-0

Weimann, Amy, Blaise Nguendo-Yongsi, Christopher Foka, Uilrich Waffo, Pamela Carbajal, Remy Sietchiping, and Tolu Oni. 2020. "Developing a Participatory Approach to Building a Coalition of Transdisciplinary Actors for Healthy Urban Planning in African Cities - a Case Study of Douala, Cameroon." *Cities & Health*, May, 1–11. https://doi.org/10.1080/23748834.2020.1741966

16 Regenerative Futures

Climate Justice and Equity in the Built Environment

Wendy Steele

Introduction: Living Cities

The future is being shaped by a rapidly urbanizing world and climate change (Solecki et al., 2024). The result is an increase in extreme heat waves, droughts, floods, storms, and catastrophic fires impacting cities and their communities (both human and nonhuman). As *living* habitats, cities and urban regions offer shelter and food for human and nonhuman communities. For those most vulnerable – the strays, the outsiders, the disaffected, and those that have been colonized, silenced, or oppressed – the built environment can be predatory and extractive. Addressing questions of climate justice and equity in the built environment is central to achieving more regenerative futures in a climate of unprecedented change.

The sheer scale and impact of the built environment on people, places, and the planet are exacerbated by unsustainable patterns of development, urban growth, production, and consumption. Whilst some people (the few) are disproportionately wealthy, for others (the majority) urban poverty and precarity has become even more entrenched. Building climate justice within complex urban histories and ecologies requires a politics of care and repair. To this end, the chapter is structured around the following three key priorities for addressing climate justice in the built environment:

1 Better recognizing living systems underpins the quest for a climate-just city.
2 The built environment offers opportunities to address and reduce climate vulnerability.
3 Transdisciplinary research is needed to achieve more regenerative futures.

By bringing together transdisciplinary research and practical examples from diverse urban contexts, this chapter draws attention to the need to prioritize climate justice in the built environment in response to a rapidly warming world. Cities and urban regions, despite all their problems, emerge as likely to be the best and safest places to be in the climate- and heat-challenged near future (Bai et al. 2018, Steele, Handmer and McShane, 2023). This comes with the need for community-led acts of urban care and repair to better support existing cities

DOI: 10.4324/9781032631110-17

and safe and just futures for people and planet. More regenerative urban futures beckon.

Revisiting the Climate-Just City

The relationship between humans and their urbanized environments has led to planetary system threats and breakdowns (Steffen et al., 2018). Through the burning of fossil fuels and other activities that feed urban progress and prosperity, the built environment has moved from the periphery to the center of the climate crisis. The 2022 Intergovernmental Panel on Climate Change (IPCC) report *Climate Change: Impacts, Adaptation and Vulnerability* makes clear that human and ecosystem vulnerability are interdependent and entangled. The growth of cities and the reduction of biodiversity-rich green spaces impact all species with extinction rates of flora and fauna unacceptably high and rising. The built environment is a critical arena for transformative change.

A climate justice approach to cities challenges the dualisms that are deployed and reinforced in urban policy and planning for the built environment around: nature–culture (ontology), present–future (epistemology), and global–local (geography) (Hulme, 2010). If climate equity "is no mere abstraction" (Low and Gleeson, 1998), then addressing the intertwined plight of both the planet and those most marginalized must be at the heart of any progressive research, policy, and community activist agenda. Such an agenda includes a deliberate focus on "who dominates?", "who benefits?", and "who gets left behind?", and how the resultant priorities then become manifest eco-socio-spatially. More specifically, this entails a focus on the following:

- How do the practices and stories of climate justice connect people and the planet?
- How does climate change exacerbate the marginalization of vulnerable communities?
- How does the climate-just city challenge, complement, or replace current rights and privileges?

Previous research on planning for climate equity and justice in the built environment with my colleagues Diana MacCallum, Jason Byrne, Donna Houston focused on the *climate-just city framework* as a lens for taking this agenda forward within the context of climate change. As a conceptual framework, the emphasis was placed on the need to integrate considerations of *social justice* (people), *environmental justice* (places), and *ecological justice* (non-humans) drawing on the insights from cultural political economy, urban ecology, the cultural politics of nature, and community development studies. The framework sought to direct research and policy attention to how many of the responses to climate change impacts in the built environment can further 'hurt people' by concentrating, entrenching, and reproducing social, spatial, and ecological injustice (see Steele et al., 2012).

The climate-just city approach emphasizes the need to attend to the complexity and connectivity of living systems as well as the stories and visions represented by different actors (human and non-human) within place-based and planetary contexts. The ground-truthing of the climate-just city framework as critical urban praxis occurs within the "situated stories and practices" involving activists, citizens, and policymakers. By incorporating diverse perspectives and adopting an open approach to their interpretation, the framework is designed to enhance our understanding of limitations and potentials for urban climate justice

Our point of departure then - as it is now - is the urgent need to find and embed ethical and socially innovative ways of enabling climate equity and justice within urban processes. This includes the need to actively engage with progressive alternatives to make more explicit and put to work the synergistic links that shape specific (in)justice practices and contexts. As prefigurative praxis this was envisaged not as a prescriptive model for policy and planning instrumentalism, but as a "process of learning involving a collective of human and more-than human actors – a process of co-transformation that re/constitutes the world" (Roelvink et al. 2009, p. 1). This is achieved by:

- Advancing our understanding of how various urban actors imagine, conceptualize, and practice climate equity and justice at various scales;
- Critically exploring how institutional and cultural perspectives shape urban imaginaries, issue-framing, and practices in response to the climate emergency; and
- Developing and applying socially innovative methods for addressing climate change in policy, planning, and community practices.

Grassroots community-based activism that supports action on climate change and promotes climate justice is already underway (Steele et al., 2023, Von Mering et al. 2025). The actions of climate groups and coalitions that have emerged in cities throughout the world emphasize how people can come together to make connections between social and environmental justice and enact meaningful solutions to climate change. These struggles lie at the heart of what we have termed the *climate-just city*.

However, in this critical climate decade to 2030, more effective strategies and frameworks for achieving climate justice in the built environment are required–particularly ones that engage directly with the practices, stories, and perspectives of climate vulnerability and those who have been traditionally excluded from planning policies and practices. Three themes emerge as critical for the future of urban climate action: decentralised and multi-actor model of transformative change; active engagement with both the local community and at the planetary-scale; and the role of transdisciplinary research that centres First Nations in this.

A Warming World: Reducing Urban Climate Vulnerability

Dominant approaches to climate vulnerability in the built environment often miss the underlying foundations that structure the unequal city in a warming world. Normative responses to the devastating scale and impact of a warming world for example are typically grounded in more techno-scientific solutions, market logics and structures, and risk management initiatives to "fix" the problem. This overlooks understanding the "who," "how," and "why" of heat-related disaster responses – and how this might change as foreshadowed in the climate-just city framework. Urban dynamics are shaped by the forces of agency and structure, power and interest, resources, skills, and capabilities. The challenges include financial arrangements, institutional structures, decision-making processes, and security discourses to support urban systems to transform towards more regenerative futures.

Building from the framework of the climate-just city, a transdisciplinary research approach to urban heat threats and disasters was undertaken by John Handmer, Ian McShane, and myself from 2020 to 2023. As three social scientist-activists and citizens working on issues of climate justice, urban governance, disaster management, science, culture, and technology, our focus was in an era of significant change in urban heat-related events and disasters, the urgent need to rethink the idea of equity in the built environment by drawing together diverse conceptual, speculative, and practical policy responses to the urban heat problem (Steele et al., 2023). Of particular interest were the interspecies, intragenerational and intergenerational justice themes of marginalization, abandonment, resistance, empathy, care, and transformation and how they impact diverse communities in cities and urban regions.

The transdisciplinary approach has gained traction in the context of addressing societal missions such as climate change by purposively including and braiding diverse knowledge systems and practices so that "no-one is left behind." However finding ways to work across very different epistemic communities is still evolving as part of the shifting landscape of knowledge generation, production, translation, and impact for transformative change (Bulkeley et al., 2023). Transdisciplinary methodologies are a response to complex contexts and the need to better attend to real-world problems that are rooted in the interactions of social, natural, and technological systems – to both unsettle and transcend traditional disciplinary silos and boundaries (Holzer, Carmon and Orenstein, 2018). The ambition has been described as "an interpenetration of epistemologies" (see Balsiger 2004) to create transformative change in, and of, the world.

> In envisioning the future of transdisciplinary initiatives.... a pressing sense of urgency emerges to address society's intricate challenges. The transdisciplinary approach becomes imperative, crucially breaking down the barriers isolating disciplines and stakeholders, fostering community resilience and social cohesion, and effectively addressing multifaceted problems.
>
> (Tapeaor 2024, p. 1)

Transdisciplinary research has also been specifically identified as helping drive a justice and equity agenda. Patricia Leavy (2016, p. 1), for example, describes the value of addressing contemporary societal challenges and social problems/

research questions from a range of critical perspectives "unhampered by the theoretical and methodological restrictions of traditional disciplinary boundaries." The research process needs to be both "educative" and "reflexive" to accommodate this fusion of disciplinary perspectives, modes of enquiry, and participatory methods (McGregor 2015).

As part of the Hot Cities transdisciplinary project, we identified 40 experts and practitioners from across the globe to make up the interplay of different voices from across diverse disciplines, practices and contexts using a snowball sampling technique. From Thailand to Togo, Australia, Europe, the UK, the USA, and Canada, these co-researchers and collaborative authors hold backgrounds in academia, activism, the public sector, and private practice and bring insights from the Humanities, Arts, Social Sciences, and Sciences in areas such as fine art, music, writing, life history, geography, climate science, philosophy, meteorology, pyro-geography, futures studies, community development, Earth-centered legal frameworks, environmental risk, and ethics, as well as human and nature rights advocates and activists.

Each of the contributors was invited to discuss a series of prompts and provocations around the theme "Hot Cities" – contexts, meanings, experiences, connections, resonances, ethics, and futures – which they were encouraged to interpret in their own way, drawing on relevant examples. These were then grouped and thematically analyzed around selected keywords (Fire, Climate, Shelter, Risk, Roots, Community, Networks, Technology, Nature, Futures, Endless Summer, etc.). What is important about the transdisciplinary approach is not just the bringing together of stories but also attentiveness to the ways in which shared stories widen and deepen understandings of both the urban heat crisis and ways to move forward. This is the active relationship of ideas and practices across the diverse life journeys and cultural and disciplinary contexts of the contributors. This is a shared keyword project and collective urban equity and justice agenda that helps to attune to new realities and dismantle the binaries of life/matter, human/animal, and organic/inorganic in the context of a climate emergency.

What emerged from the research was the challenges and opportunities of understanding the built environment as a series of nested and interconnected living systems in a warming world: where human and other life and materials are mutually embedded and increasingly threatened. This is resonant with alternative economist Kate Raworth's calls for radical transformations to society to support safe planetary thresholds that are both regenerative and distributive. In her Doughnut Economics model, she aligns the planetary boundaries necessary for the survival of life on Earth with the need for a socially just space within which humanity can survive and thrive (Raworth, 2017). This equity and justice focus emplaces the critical dimensions of the built environment (e.g. food, water, energy, networks, housing) within the complex interrelationships and systems involved in planetary-scale climate change (see Figure 16.1).

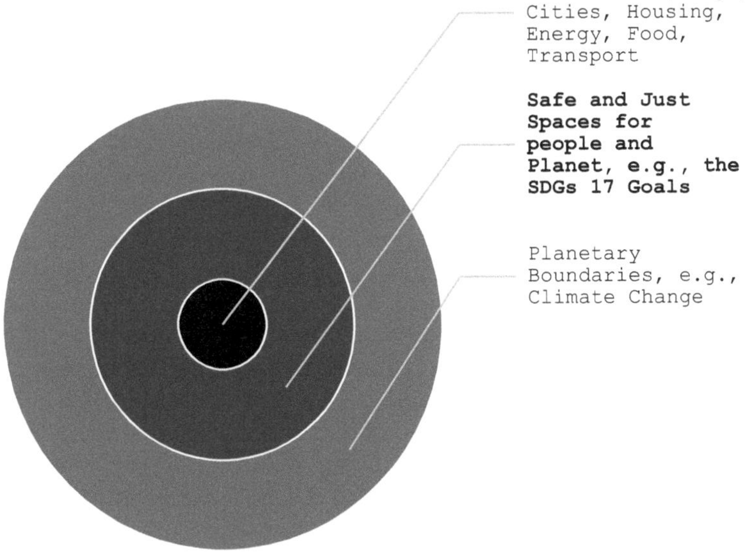

Figure 16.1 Safe and just spaces in the built environment. Adapted from Raworth (2017).

The framework offers a compass for taking regenerative action and is designed around creating a safe and just space for humanity to flourish. As with the climate-just city framework, it is not designed to offer prescriptive recommendations but acts as a heuristic guide for local-scale innovation and participatory climate action in the community sector, not-for-profits', and progressive private industry, universities, and cities through the development of city portraits. The *Doughnut Barcelona*, for example, offers an ecological and social approach that seeks to link the well-being of the planet to flourishment at the city scale (see Barcelona City Council, 2024).

The City of Barcelona developed a comprehensive review of key areas in the built environment where action is most urgently needed and making these transparent to citizens, policymakers, and innovators as a collaborative civic agenda: urban greening, buildings and housing, waste, water and energy, critical infrastructure, etc. The ambition was to help Barcelona "build a more sustainable, equitable, and prosperous city that respects both local and global ecological boundaries while ensuring social justice and wellbeing for its citizens" (Cattaneo 2023, p. 1). As the Barcelona example highlighted, "The challenge Barcelona faces is multifaceted: how to boost the city's social foundation while not further infringing upon biophysical boundaries, and to cut back current oversteps. A daunting task, indeed, but one the city cannot afford to ignore" (Cattaneo 2023, p. 2).

Toward Regenerative Urban Futures

The climate crisis in the context of the built environment brings into conversation very different storylines around "the nature of human nature." Urban climate justice is about reducing climate vulnerability in cities and developing critical dialogue and ethical practices in a shared ecological emergency where our "strength is not as an individual but as a collective" (Thunberg, 2019, p. 15). This also involves the mobilization of alternative knowledge including the vital role of Indigenous cosmologies, as well as radically different visions and experimental practices taking root at the grassroots within local environments. An equitable built environment in the context of urban vulnerability and precarity involves an expanded imagination to better respond to the impacts of climate change.

The range of ways of "knowing" cities in the context of climate change includes both the informal and subjective such as tacit and personal experience, local knowledge, visual images, actions, and activities, as well as the more formal economic and bureaucratic reports, resource mapping, and scientific research. The purpose of bringing together shared stories is to tether together these intimate insights, bound up with "organising hope, negotiating fears and mediating memories" (Sandercock, 2002, p 4.). Stories are regenerative in this sense, and we draw on different knowledge practices "to see, think and feel anew… 'anew' in a generative sense" (Armstrong, 2008, p. 20).

In the following section, four key themes to emerge from the *Hot Cities* transdisciplinary project are summarized below, guided by the principles of the climate-just city framework and the regenerative futures of safe and just societies (Figure 16.1). The paradox that is highlighted is that the pursuit of human development has the potential to undermine the Earth-system processes upon which it depends. The nested nature of interconnected living systems underpins the built environment – the ways in which human, other species and planetary precarity are deeply connected. This in turn foreshadows the critical role of sharing the transdisciplinary stories/practices needed for transformative change, and the implications of this for climate justice and equity in the built environment.

1 *Transformative change in the built environment is possible*

There are many positive examples of change that support urban climate justice in practice – from ways to reduce heat and stay cool at a citywide scale to rethinking policy, planning, and institutional design to better address questions of risk and uncertainty, as well as different ways of being that engender more equitable and inclusive planetary futures. There is a growing understanding that the Earth itself is alive and vulnerable to extractive human overactivity and development. Humanity's relationship with the planet comes with responsibilities of care and repair that we pass to our children and all forms of future life.

Indigenous and First Nations knowledges offer Earth-centered and cosmological visions of community as an intimate interaction between the mind and spirit (abstract world), with the land, relationships, and activity (physical worlds) (Yunkaporta, 2019). This is a radically inclusive notion of life and well-being that is dynamic and relationally interconnected to all more-than-human entities, including land, rivers, weather, trees, birds, fish, and humans (Dionisio et al., 2022). As Yunkaporta describes, "creation" is about what we collectively create through the sharing of cultural practices.

Decolonial and degrowth manifestos from around the world have drawn attention to the need to recognize the roots of crisis linked to colonization, carbon, and capitalism through a shift toward the ecological and social commons. Environments, especially urban environments and the landscapes around and in cities, should, and in some locations are, be restored and regenerated by communities. In cases highlighted by the research such as in New York City and Melbourne, this involves greening everything possible, as part of a kaleidoscope of urban activity with water, plants, and endemic species as key priorities (see Grabowski, McPhearson and Pickett, 2023). This is closely related to increased recognition of the vital potential of greenery for cooling the built environment as well as other reciprocal benefits of eco-social commons as care infrastructures for both humanity and the living planet (Mellick-Lopes et al., 2018).

Cities large and small are on the way to being fully renewable in terms of energy; Canberra is one example where the transition to renewable energy systems is well underway. Shanghai has been very successful in the widespread uptake of electric vehicles and city-states like Singapore are leaders in urban greening for heat reduction. Athens has developed and is implementing both short and long-term strategies for heat-related sustainability. But much if not most of what is written about cities is set within the formal, official, municipal, and government settings. Between a third and a quarter of the urban world occupies precarious informal settlements where risk is often exacerbated by climate displacement, poverty, and exclusion (Zapata Campos et al., 2022). Better understanding informality in cities is critical to resilience and the prospects for "just sustainability" (see Agyeman, Bullard and Evans, 2003) in a rapidly urbanizing world.

2 *Climate justice underpins urban policy, planning, and decision-making*

Social and ecological tipping points are driving action on climate change. Ensuring justice and equity across humans, nonhumans, and the biosphere is the preeminent requirement for livable future cities – and is how we conceptualize justice and equity. It is a moral obligation to do everything we can to make sure the planet is inhabitable for not only those humans coming after us but also the rest of life on the planet as well (Steffen 2018). This will need a change in power structures and a much stronger focus on ethics and justice for all living things. There is now much discussion and consideration of these issues as part of the climate crisis. The impacts on people, their livelihoods, and things they value vary enormously – although all are affected.

The transformation of discriminatory or racialized urban environments is linked to questions of climate justice including sovereignty and sea-level rise negotiations, advancements on treaties and the impact of climate change on stolen land, the redistribution of resources and funds to create cooler living environments, experiments with universal basic wages, universal insurance for natural hazards, as well as future city redesign and relocation preparations. These can involve trade-offs between welfare and liberties, although this raises moral questions and decisions around whose welfare and what sort of liberties. Recognizing the history of development trajectories and impacts matters. History cannot be erased and must be meaningfully addressed as part of the process of reimagining sustainable urban futures.

Cities are more vulnerable due to their heat islands, and health and safety are major issues as heat can prevent essential work from occurring. But not all cities are equal and not all who live in cities are counted as active citizens, with many on the fringe and living very precariously, at times surrounded by enclaves of extreme wealth and prosperity. No one is protected from the climate crisis, but risk amplifies the inequities and vulnerabilities in our societies and cities. Some are more impacted by their color, geography, species or gender, or by the actions of others beyond their control, far more than others are. The focus on ethics and equity is at the forefront of the needed transitions to more sustainable cities, including the approaches taken to cooling.

The "just transitions" concept, (see Just Transition Centre, 2017) developed by US labor unions and originally about the transition away from coal before broadening to encompass systemic injustice provide guides. This includes support for First Nations people and their ongoing struggles on stolen, unceded land, as well as building a social movement of change toward an ecologically stable and more equitable society. Those dependent on the climate for their livelihoods or security are likely to be the worst affected. Chronic food shortages and housing and employment insecurity all impact climate-precarious communities.

Technology is stepping into the transition space (Strengers and Kennedy, 2021). It is a large part of the future, and of taking action on climate change. Good models, standards, and approaches for managing the downsides of technology, in particular AI, exist and communities could in many countries like Australia relatively easily and simply adopt them. Technologies are part of urban futures whatever pathways are taken. Changes in physical infrastructure will also be needed to mitigate and adapt to climate change, but complementary social and behavioral changes are needed as well. For this, sociotechnological futures need reimagining – by government and civil society. The promise and ethics of intelligent technologies must be matched by a social contract of civic care and moral responsibility (Steele and Stefánsson, 2021).

3 *Reshaping our institutions to reduce inequalities*

The present political system is unfit for the problems we face including that of justice and equity for all elements of the planet and the role of the built environment in addressing this. We need (much) stronger institutions to solve the problems of inequalities, climate change, and sustainability – among others. There is consensus among the contributors we interviewed that we are not going to solve global warming through individual actions or tribalism which risks isolation. Instead, we need better leadership, coordination and collaboration, and the co-creation of community action at all scales through the strengthening of social ties. This means building stronger connections between living cities and their inhabitants, creating systems of mutual support for those most at risk, and ensuring that states do more to ward off disasters.

The biggest issues with achieving equity and justice in the built environment are the power structures and the decision-making mechanisms that shape who is allowed to make decisions, about what issues, and to what extent their participation has influence on decisions. A key question for those making decisions on behalf of society (others) is whether the decisions should attempt to be risk neutral. There are important trade-offs between making decisions with speed and democratic processes. One view is that decisions should be precautionary and therefore risk averse especially where inequalities and climate are concerned. Managing risk through incremental improvements against exponential change rather than step change is the status quo. In a climate emergency, there are calls for more radical shifts to address the crisis – a metamorphosis of society and carbon-intensive development activities (Latour, 2021).

Despite the science of climate change showing very clearly what happens in a hotter world, we continue to face dangerous predatory delays by those who have a greater interest in slowing things down than speeding things up. Urban leadership and policy captured by vested interests in the fossil fuel, transport, and construction sectors has been identified as a critical blockage to clean energy transitions or in addressing poverty, species extinction, and environmental degradation. This exacerbates the distributional impacts of the built environment and increases the precarity and vulnerability of those least able to cope with the impacts of climate change.

Addressing this means more clearly understanding the various ways delay shapes policy processes; the demand for an orderly transition can too often be an excuse to slow down what needs to be rapid change. Accelerating the process of mitigation and decarbonization, in the context of institutional inertia, raises important questions about civic participation and the extent to which this can challenge dominant political and institutional trajectories and extractive mindsets. Climate action occurs at the intersection of eco-social justice, ethical decision-making, and governance of the built environment. Civil society has an enormous, largely untapped creative capacity and desire to contribute to change leveraging political pressure points in key areas such as institutional integrity, affordable housing, renewable energy, and new forms of collective action needed to address the biodiversity and climate – but also ethical and spiritual – crisis.

4 *Reclaiming hope for regenerative futures*

Cities and the urban hinterlands that support them are diverse and dynamic, generators of opportunities as well as all the problems and issues they embody and amplify. We cannot go back to what was there – the world, environment, and cities themselves have changed and continue to evolve away from whatever was there before. Emotionally, we need courage, to take chances and invest heavily in ways that give us the best possibility of a safer future. A reminder that the challenges of the built environment in the face of climate change can be bleak, but there are also practices and pathways that hold kinship and beauty and offer ways of finding meaning and purpose in dark moments.

The arts and creative practices can transform individuals, communities, spaces, and places and this comes in many forms. Creative work – whether real or fictional – helps us comprehend the impact of the climate crisis, where every interpretive act is a recreation, a source of hope, motivation, reflection, and renewal. Design, storytelling, science, sociology, music, philosophy, history, art, ecology, planning, and community activism all have the capacity to reframe our thinking and influence us to see the world in new ways. This includes the power to captivate and engage the senses, raise and respond to questions, inspire empathy, generate thought, and encourage and mobilize climate action.

For many the feelings of devastation and loss can be overwhelming in the face of urban heat threats and disasters, or more systematically in the face of ongoing racism and land theft, global poverty and inequality, and the widening gap between those with the resources and technology to adapt to the climate crisis and those without. Existentially, this is a shared global challenge of meeting carbon emission targets to sustain life on the planet, requiring humanity to cooperate in ways it has never done. Yet, at the same time, as with the experience of urban fires, events such as this can also lead to pathways of recovery and renewal and the creative exploration of radical possibilities.

Responding to the climate crisis requires creative approaches to complex situations that go beyond scientific language that is critical but may not be easily understood. Thomas Berry (1988) wrote that a crisis is when the story of change becomes inadequate for meeting the demands of the present. The use of different storied ways and different languages helps to make difficult concepts, decisions, or scenarios more accessible. This in turn can help communities to see change, share fears of losing everything, and rethink what is important, what is possible, and how climate justice in the built environment might be reimagined in the quest for more regenerative futures. International climate change champions Christina Figueres and Tom Rivetto Carnac (2020) highlight this in their experience with the global negotiations that led to the Paris Agreement, a legally binding international agreement on climate change:

> We will not have a regenerative Anthropocene by default, but we can create it by design. With directional intent we can shift our aspirations from our current extractive growth to a life sustaining society of regenerative values, principles and practices.

Conclusion: Taking Climate Action When Systems Break Down

Achieving urban justice and equity and reducing climate vulnerability in the built environment is critical in a rapidly warming world so that both people and the planet can flourish. This requires more discerning choices and identifying ethically informed points of connection that link environmental, ecological, and social concerns to create more sustainable and regenerative futures. Better understanding, mapping, and acting on the challenges faced by those already impoverished and suffering hardship within our societies within the context of the climate emergency. This includes the need to create different political–ecological practices through more engaged urban citizenship. The acknowledgment and removal of large-scale injustices such as those experienced by First Nations people is a key part of confronting the systemic inequities that underpin the development and expansion of cities.

Transdisciplinary stories of regenerative change and possibility have a critical role that can work to "unsettle the settled," help mobilize community, or generate empathy in climate change. The politics and poetics of storytelling offer creative ways for sharing diverse experiences, navigating the complexity of how different realities and futures are being constituted and reconstituted; particularly for young people confronting the climate crisis where climate anxiety has been rising including feelings of anxiety, grief, and hopelessness when thinking about the future (Verlie, 2021).

Regenerative urban change requires recognizing the deep, living systems interdependencies that exist on Earth, as well as taking steps to reset and reframe urban practices to be more just and equitable in the built environment. If we are to understand and prepare for the impact of climate threats and disasters, greater understanding of urban and ecological processes is critical – especially when human and planetary systems break down.

Acknowledgments

"Planning the Climate Just City" was published in *International Planning Studies* in 2012 with co-authors Diana MacCallum, Jason Byrne, and Doona Houston. The "Hot Cities" research project is a project co-led by John Handmer and Ian McShane who were involved in the design, analysis, and writing of the book *Hot Cities: A Transdisciplinary Agenda* published by

Edward Elgar in 2023 which includes the expert input of international contributing authors: Cathy Applegate, Xuemei Bai, Christian Barry, David Bowman, David Carlin, Danielle Celermajer, Mark Crosweller, Niki Frantzeskaki, Tony Fry, Isabella Gerometta, Jody Graham, Stephen Healy, Jean Hillier, Simon Kerr, Eric Klinenberg, Jo Lane, Crystal Legacy, Michelle Maloney, Simon Marvin, Darryn McEvoy, Timon McPhearson, Abby Mellick Lopes, Therese Milanovic, Eleni Myrivili, John Nairn, Alan Pears, Sarah Pink, Libby Porter, Stephen Pyne, Lauren Rickards, Kaossara Sani, Wendy Sarkissian, Benedict Sibley, Katie Steele, Will Steffen, Yolande Strengers, Pakamas Thinphanga, Blair Trewin, and Cam Walker.

Bibliography

Aflaki, A., Mirnezhad, M., Ghaffarianhoseini, A., Ghaffarianhoseini, A., Omrany, H., Wang, Z., and Akbari, H. (2017) Urban Heat Island Mitigation Strategies: A State-of-the-art Review on Kuala Lumpur, Singapore and Hong Kong, *Cities*, 62, p. 131–145.

Agyeman, J., Bullard, R., Evans, B. (2003) *Just Sustainabilities*, London, Earthscan.

Armstrong, C. (2008) *Women Artists at the Millennium*, Cambridge, MIT Press.

Bai, X., Dawson, R. J., Ürge-Vorsatz, D., Delgado, G. C., Barau, A. S., Dhakal, S., Dodman, D., Leonardsen, L., Masson-Delmotte, V., Roberts, D., Schultz, S.. 2018, Six Research Priorities for Cities and Climate Change, *Nature*, 555 (7694):23–25.

Balsiger, T. (2004) Supradisciplinary Research Practices: History, *Objectives and Rationale, Futures*, 36 (4), p. 407–421.

Barcelona City Council (2024), *Barcelona for Climate*, accessed on https://www.barcelona.cat/barcelona-pel-clima/en/moves-towards-a-new-economic-model/what-is-doughnut-economics, 16th July.

Berry, T. (1988) *The Dream of the Earth*, San Francisco, Sierra Club.

Bulkeley, H., Lecavalier, E. and Basta, C. (2023) Transformation Through Transdisciplinary Practice: Cultivating New Lines of Sight for Urban Transformation, *Local Environment*, 28(7), p. 829–836.

Byrne, J., Maccallum, D., Steele, W., Houston, D. (2011) *Australia Fair: Imagining and Enacting Urban 'Climate-justice'*, Urban Research Program, Griffith University. Research paper series.

Cattaneo, C. (2023) *The Doughnut Journey in Barcelona - the process*, Barcelona, Doughnut Economics Action Lab, p. 1–2.

City of Melbourne (2017) *Nature in the City Strategy*. Accessed on https://www.melbourne.vic.gov.au/community/greening-the-city/urban-nature/Pages/nature-in-the-city-strategy.aspx

Dionisio, R., Dombroski, K., & Yates, A. (2022) Testing Practices for Testing Times: Exploring Indigenous-Led Governance. *Dialogues in Human Geography*, 13(2), p. 268–287.

Dodds, K., Castán Broto, V. et al (2022) 10th anniversary of Territory, Politics, Governance: Achievements and Prospects, *Territory, Politics, Governance*, 10 (2), p.145–158.

Elmqvist, T., Bai, X., Frantzeskaki, N. et al. (2018) *Urban Planet: Knowledge Towards Sustainable Cities*, Cambridge, Cambridge University Press.

Figueres, C. and Rivett-Carnac, T. (2020) *The Future we Choose: Surviving the Climate Crisis*, London, Manilla Press.

Frantzeskaki, N., Broto, V. C., Coenen, L., & Loorbach, D. (2017). Urban Sustainability Transitions: The Dynamics and Opportunities of Sustainability Transitions in Cities.

In Frantzeskaki, N., Broto, V. C., Coenen, L., & Loorbach, D. (Eds.) *Urban Sustainability Transitions* (pp. 1–20). New York, Routledge.
Grabowski, Z. J., McPhearson, T., & Pickett, S. T. A. (2023). Transforming US Urban Green Infrastructure Planning to Address Equity. *Landscape and Urban Planning*, 229, 104591.
Harris, F. and Lyon, F. (2014) Transdisciplinary Environmental Research: A Review of Approaches to Knowledge Co-production, *Nexus Network* 2 (2014), p. 27.
Head, L. (2016) *Hope and Grief in the Anthropocene: Re-conceptualising Human–Nature Relations*, London, Routledge.
Holzer, J., Carmon, M. and Orenstein, D. (2018) A Methodology for Evaluating Transdisciplinary Research on Coupled Socio-Ecological Systems, *Ecological Indicators*, 85, p. 808–819.
Hulme, M. (2010) Cosmopolitan Climates: Hybridity, Foresight and Meaning. *Theory, Culture and Society*, 27 (2), p. 267–276.
IPCC (2022), *Impacts, Adaptation and Vulnerability. Contribution of 511 Working Group II to the Sixth Assessment Report of the Intergovernmental Panel on Climate Change* [H.-O. Pörtner, D. C. Roberts, M. Tignor, E. S. Poloczanska, K. Mintenbeck, A. Alegría, M. Craig, S. Langsdorf, S. Löschke, V. Möller, A. Okem, B. Rama (eds.)]. Cambridge University Press.
Jackson, S. (2013) Weather of Mass Destruction: Forecasting the Future as Climate-Driven War, in Simon-Lopez, A. and Yeandle, H. (eds.) *A Critical Approach to the Apocalypse*, Leiden, Brill, p. 71–85.
Johnson, O. and Karlberg, L. (2017) Co-exploring the Water-energy-food Nexus: Facilitating Dialogue Through Participatory Scenario Building, *Frontiers in Environmental Science* 5, p. 1–12.
Just Transition Centre (2017) *Just Transition*: A Report for the OECD. Accessed on https://www.oecd.org/env/cc/g20-climate/collapsecontents/Just-Transition-Centre-report-justtransition.pdf
Knapp C., Reid R., Fernández-Giménez M., Klein J., and Galvin K. (2019). Placing Transdisciplinarity in Context: A Review of Approaches to Connect Scholars, Society and Action. *Sustainability*, 11(18), p. 4899.
Latour, B. (2021) *After Lockdown: A Metamorphosis*, London, Polity Press.
Leavy, P. (2016) *Essentials of Transdisciplinary Research Using Problem-Centered Methodologies*, New York, Routledge.
Low, N. and Gleeson, B. 1998. *Justice, Society and Nature: An Exploration of Political Ecology*, London: Routledge.
Luque-Ayala, A., Marvin, S. & Bulkeley, H. (Eds.) (2018) *Rethinking Urban Transitions: Politics in the Low Carbon City*, London, Routledge.
McGregor, S. (2015) Transdisciplinary Knowledge Creation, in Gibbs, B. (ed) *Transdisciplinary professional learning and practice*, New York, Springer, p.9–24.
Mellick-Lopes, A., Healey, S., Power, E., Crabtree, L., Gibson, K. (2018) Infrastructures of Care: Opening up "Home" as Commons in a Hot City, in *Human Ecology Review*, 24(2), Special Issue: Addressing the Great Indoors—A Transdisciplinary Conversation, p. 41–60.
Nixon, R. (2011) *Slow Violence and the Environmentalism of the Poor*, Cambridge, Harvard University.
Raworth, K. (2017) *Doughnut Economics: Seven Ways to Think Like a 21-st Century Economist*, Vermont, Chelsea Green Publishing.
Roelvink, G., Gibson, K., & Graham, J. (2009). A Postcapitalist Politics of Dwelling: Ecological Humanities and Community Economies in Conversation. *Australian Humanities Review*, 46, p. 145–158.
Roelvink, G. and Gibson-Graham, J. K. (2010) An Economic Ethics for the Anthropocene *Antipode*, 41 (1): 320–346

Sandercock, L. (2002) Practicing Utopia: Sustaining Cities, *The Planning Review*, 38(148), p. 4–9
Solecki, W., Roberts, D., Seto, K. (2024) Strategies to Improve the Impact of the IPCC Special Report on Climate Change and Cities, *Nature Climate Change*, 14(7), p. 685–691.
Steele, K. and Stefánsson, O. (2021) *Beyond Uncertainty: Reasoning with Unknown Possibilities*, Cambridge, Cambridge University Press.
Steele, W. (2020) *Planning Wild Cities: Human-Nature Relationships in the Urban Age*, London, Routledge, 580.
Steele, W., Handmer, J., McShane, I. (2023) *Hot Cities: A Transdisciplinary Agenda*, London, Edward Elgar Publishing.
Steele, W., MacCallum, D., Byrne, J. and Houston, D. (2012) Planning the Climate Just City, *International Planning Studies*, 17 (1), p. 67–83.
Steffen, W., Rockström, J., Richardson, K. and Schellnhuber, H. (2018) Trajectories of the Earth System in the Anthropocene. In *Perspective: Earth, Atmospheric and Planetary Sciences*, 115 (33), p. 8252–8259.
Steffensen, V. (2020) *Fire Country: How Indigenous Fire Management Could Help Save Australia*, Melbourne, Hardie Grant Publishing.
Strengers, Y. & Kennedy, J. (2021) *The Smart Wife: Why Siri, Alexa, and Other Smart Home Devices Need a Feminist Reboot*, Cambridge MA, MIT Press.
Tapeaor, R. (2024) *Co-Designing for a Transdisciplinary Vision*, Stockholm, Future Earth.
Thunberg, G. (2019) *No One is Too Small to Make a Difference*, London, Penguin.
Vahanvati, M., McEvoy, D., Kuh, D., Iyer-Raniga, U. (2022) *Inclusive and Disaster Resilient Shelter 604 Guide: Urban Informal Settlements, Honiara, Solomon Islands*, Suva, Habitat for Humanity.
Verlie, B. (2021) *Learning to Live with Climate Change: From Anxiety to Transformation*, London, Routledge.
Von Merring, S., Bell, T., da Silva Faustina, A., Steele, W., Ward, A., Soboron, M. (2025) *Routledge Handbook of Grassroots Climate Activism*, New York, Routledge.
Yunkaporta, T. (2019), *Sand Talk: How Indigenous Thinking Can Save the World*, Melbourne, Text Publishing.
Zapata Campos, M., Kain, J., Oloko, M., Scheinsohn, M., Stenberg, J. and Zapata, P. (2022) Residents Collective Strategies of Resistance in Global South Cities Informal Settlements: Space, Scale and Knowledge. *Cities*. 125, 103663.

Index

Pages in *italics* refer to figures, pages in **bold** refer to tables, and pages followed by n refer to notes.

MIX

Papier | Fördert gute Waldnutzung

FSC® C083411

Kolibri 360 GmbH
Ferdinand-Jühlke-Straße 7
99095 Erfurt, Deutschland
produktsicherheit@kolibri360.de